Renewable Energy

*Edited by Roland Wengenmayr
and Thomas Bührke*

Related Titles

Kruger, P.

Alternative Energy Resources
The Quest for Sustainable Energy

272 pages
2006
Hardcover
ISBN: 978-0-471-77208-8

Würfel, P.

Physics of Solar Cells
From Principles to New Concepts

198 pages with 111 figures and 7 tables
2005
Hardcover
ISBN: 978-3-527-40428-5

Luque, A., Hegedus, S. (eds.)

Handbook of Photovoltaic Science and Engineering

approx. 1168 pages
2003
Hardcover
ISBN: 978-0-471-49196-5

Markvart, T.

Solar Electricity, 2nd Edition

298 pages
2000
Hardcover
ISBN: 978-0-471-98852-6

Renewable Energy

Sustainable Energy Concepts for the Future

Edited by
Roland Wengenmayr and Thomas Bührke

WILEY-VCH Verlag GmbH & Co. KGaA

The Editors

Roland Wengenmayr
Frankfurt/Main, Germany

Thomas Bührke
Schwetzingen, Germany

German Edition translated by:
Prof. William Brewer

Additional articles translated by:
C. Benjamin Nakhosteen

Library of Congress Card No.:
applied for

British Library Cataloguing-in-Publication Data
A catalogue record for this book is available from the British Library.

Bibliographic information published by the Deutsche Nationalbibliothek
Die Deutsche Nationalbibliothek lists this publication in the Deutsche Nationalbibliografie; detailed bibliographic data are available in the Internet at <http://dnb.d-nb.de>.

© 2008 WILEY-VCH Verlag GmbH & Co. KGaA, Weinheim

Typesetting   TypoDesign Hecker GmbH, Leimen

Printing   betz-druck GmbH, Darmstadt

Binding   Litges & Dopf Buchbinderei GmbH, Heppenheim

Cover Design   Adam-Design, Bernd Adam, Weinheim

Printed in the Federal Republic of Germany
Printed on acid-free paper

ISBN   978-3-527-40804-7

Preface

The imminent threat of catastrophic climate change caused by anthropogenic, i.e. man-made changes of the composition of our athmosphere, especially the concentrations of carbon dioxide (CO_2), laughing gas (dinitrogen monoxide, N_2O) and methane (CH_4), has been widely recognized. The CO_2 concentration is today worldwide above 380 ppm, far above the maximum CO_2 concentration of about 290 ppm observed in the last 800,000 years. The recent report of the Intergovernmental Panel on Climate Change (IPCC) and the agreements of the Bali Climate Summit in December of last year demonstrate that the world starts to face the technological and political challenge poised by the requirement to reduce the emission of these gases by 80% in the next few decades.

This goal can only be achieved by substantial progress in the two big areas that affect this issue: Rapidly enhanced production of energy from renewable sources and increased energy efficiency, especially of buildings where a large amount of our total energy need is generated.

This book addresses both of these critical objectives with 19 contributions written by experts in their respective fields, covering the most important issues and technologies needed to reach these dual goals. This volume provides an excellent, concise overview picture of this important area, combined with interesting details for each topic for the specialists. The topics addressed include photovoltaics, solar thermal energy, geothermal energy, energy from wind, waves, conventional hydroenergy, bioenergy, hydrogen technology with fuel cells, building efficiency and solar cooling. In each chapter, the detailed discussion and references to current literature enable the reader to reach an own opinion concerning the feasibility and potential of these technologies. The volume appears to be well suited for the generally interested reader, but may be well used in advanced graduate classes on renewable energy. It seems to be especially well suited to assist students who are in the process of selecting an inspiring, relevant topic for their studies and later their thesis research.

Eicke R. Weber,
Director,
ISE Institute for Solar Energy Systems,
Freiburg, Germany

Renewable Energy. Edited by R. Wengenmayr, Th. Bührke. Copyright © 2008 WILEY-VCH Verlag GmbH & Co. KGaA, Weinheim. ISBN 978-3-527-40804-7

High-quality, First-hand Information

Renewable energy is a key concept for the 21st century. No other area of technology is accompanied by so much optimism and hope that humanity can meet the challenges of climate change and a secure energy supply in an intelligent manner. The increasing number of wind power plants, solar collectors and photovoltaic installations demonstrates perceptibly that many innovations for tapping renewable energy sources have matured. Hydroelectric plants with their dams and reservoirs have long since been a part of the landscape; in fact, we have them to thank for the second industrial revolution of widely-available electrical energy grids. Other technologies, such as the use of geothermal heat, which is abundantly available almost everywhere, are entering the stage of large-scale pilot projects. Very few technologies have developed so dynamically in the past years as have renewable and alternative energy supplies.

This book gives a detailed overview of the current state of the most important technologies which are already contributing significantly to our energy supplies – or will be able to do so in the future. Each technology is described by authors who as researchers, engineers, or entrepreneurs are experts in their fields, which makes this book an especially valuable and reliable source of information. Here, well-grounded explanations are given of e.g. how a solar-thermal power plant operates, or which new developments promise to solve the serious cost and energy problems of the established silicon technology for photovoltaic cells.

The book not only introduces technological methods for obtaining energy from renewable sources. It also contains important information about how energy can be efficiently stored, transported and converted to useful forms. How heavy must a battery be in order to store the same amount of energy as a tank full of hydrogen? Which pressing problems could be realistically solved by a hydrogen economy? How does a fuel cell work? These and many other questions are answered by our authors.

When we search for possibilities to allow us to make more intelligent use of the valuable energy resources, then an important human habitat comes into view: our dwellings. In particular, very large buildings where a large number of people live, work and do their business offer an enormous potential for saving energy – and for reducing emissions of greenhouse gases – by applying intelligent architectural and air conditioning concepts. Air conditioning counts worldwide among the notorious energy gluttons, and its market is practically exploding in the warm emerging nations. Thus, the question is becoming more and more pressing as to how to provide air conditioning for buildings in an environmentally-friendly and energy-conserving manner. There are fascinating answers to this question; even the Sun itself can be tapped for cooling. How that works is explained in this book, which also introduces the world's first and thus far only high-rise building to be air conditioned in an environmentally favorable and energy-saving way by applying intelligent ideas.

The attentive reader will notice that there are only two contributions to the subject of biofuels in this book. For several reasons, we have refrained from including more on this subject. For one thing, there are currently established technologies for producing biofuels which are, from the technical point of view, not particularly new or exciting. Secondly, the ecological balance of the bioethanol and biodiesel fuels – so-called first generation agro fuels – produced today must be considered critically: their cultivation, transport and processing consume a large amount of valuable land and water, and are anything but climatically neutral. But there are good ideas in this area as well for future second generation agro fuels, which can solve these problems. A patented high-tech process for converting plant waste into valuable energy carriers which promises a very positive energy and environmental balance is introduced in this book.

The authors of this book are from Germany, since the German Federal government has massively subsidized renewable energy technologies for many years, more than any other industrial nation thus far. Thanks to this cutting-edge role, German scientists, engineers and companies have conquered a leading position worldwide in many areas of renewable energy sources, for example in wind power and photovoltaic conversion.

All the contributions are written in a readily understandable style; readers with a general educational background in technology and natural sciences will be able to follow them without difficulty. Only a few articles include some mathematics for those who wish to penetrate the subject more deeply. These few, brief passages can be simply skipped over if desired, without losing the thread of the discussion. Extensive literature lists and Web links offer many possibilities for delving further into the topics.

All of the numbers and facts have been carefully checked, which is not to be taken for granted. Precisely in the area of renewable energy sources, there is much misinformation and misleading talk in circulation. Therefore, this book intends to offer to all those who are interested a reliable, solid base of information, and to be useful also as a reference work. Whoever reads it carefully will be able to discuss the subject competently, and in particular to make informed decisions.

We thank all the authors for their excellent cooperation, Dr. William Brewer for his careful translation, and the publishers for this beautifully designed and colorful book. In particular we thank Dr. Ulrike Fuchs and Nina Stadthaus of Wiley-VCH Berlin for their skill, support and patience with us. Without Ulrike Fuchs's commitment this wonderful book might have never been realized.

Roland Wengenmayr and Thomas Bührke

Frankfurt am Main and Schwetzingen, Spring 2008.

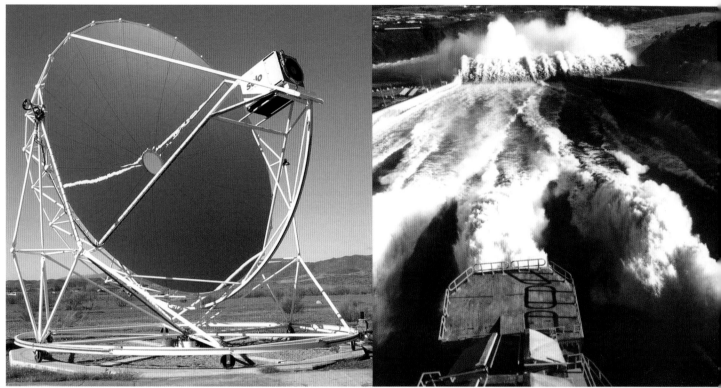

Photo: DLR   Photo: Voith Siemens Hydro

Contents

Renewable Energy. Edited by R. Wengenmayr, Th. Bührke. Copyright © 2008 WILEY-VCH Verlag GmbH & Co. KGaA, Weinheim. ISBN 978-3-527-40804-7

Photo: Vestas Central Europe

Photo: GFZ

This large photovoltaic roof installation, above the Munich Fairgrounds building, has a nominal power output of about 1 MWe l. In 2004, it fed around 1000 MWh of electrical energy into the power grid (Photo: Shell Solar).

The Development of Renewable Energy Carriers

Renewable Energy Sources on the Rise

BY HARALD KOHL

The threat of a catastrophic climate change and rising petroleum prices are putting renewable energy sources at the center of public interest. How large is their contribution at present in the world? How great is their potential for expansion? A progress report and a glance at Germany whose Federal Government has for years strongly promoted renewable energies.

Renewable energy is developing towards a recipe for success, in Europe, the USA, and Asia. As an example, let us start with taking a more detailed look at the political developments in the European Union. Here, it is chiefly the so-called "20-20-20" resolution of the European Council of March 2007, which has strongly promoted of renewable energy. The European heads of states and governments declared in it, among other things, that the share of renewable energy in the total energy consumption of the EU must increase to 20%. Today, it is 6.6%. This goal is to be achieved in equal mandatory national shares. For biofuel, additionally, a binding minimum goal of 10% was specified as the fraction of biogenic fuel in the total petrol and diesel consumption for all EU member states until 2020.

A further step will involve a comprehensive directive on renewable energy in the European Union. The first draft was announced for the second half of 2007. It will supplement the 2001 EU directive on regenerative electrical power production and will in future cover all areas of renewable electrical power, heat, and mobility.

Today's image of renewable energy usage in Europe is still dominated by a more traditional view. Sweden, Finland, and Austria, in particular, feature shares of 20 to 30 percent renewable energy out of the total primary energy consumption, due to high amounts of electricity from hydropower. Considering heat consumption, France also provides a substantial portion from renewable sources, mainly from biomass. Solar thermal usage is widespread particularly in Germany, Austria, and Greece. Markedly, conditions

Renewable Energy. Edited by R. Wengenmayr, Th. Bührke. Copyright © 2008 WILEY-VCH Verlag GmbH & Co. KGaA, Weinheim. ISBN 978-3-527-40804-7

An offshore windpark on the high sea might look like this.
(Graphics: Nordex.)

in the sunny South of the European Union vary significantly. Whereas approximately three million square meters of solar thermal collectors are installed in Greece, their number in Italy, Spain, and Portugal adds up to only a few hundred thousand square metres. Photovoltaics also are by no means a Mediterranean specialty. Here, Germany is far ahead with 1 910 000 kWp (kilowatt peak power), but Italy, in terms of kWp peak power, is even behind the not exactly sun-blessed Netherlands with 50 776 kWp.

Nevertheless, the rise of renewable energy is remarkably dynamic in many European countries. Belgium, for example, even though at a low level, has nearly tripled its share of renewable energy between 1997 and 2005, from 1.0 to 2.8 percent. Strong rises are discernible in the central and eastern European acceding countries too, such as Hungary and the Czech Republic, but here also in the low one-digit percentages. Thus, many countries will have to make considerable efforts to achieve the objectives for 2020.

Wind Energy is Booming in the USA too

Wind energy is a particularly good example showing the different degrees of success even under comparable initial situations. The general conditions of energy policy are vital here. The German Renewable Energy Sources Act (EEG) with its investment-friendly delivery and compensation regulations – together with the similar Spanish legislation – compares favorably on an international scale. 18 EU Member States have to date taken over such primacy and delivery regulations. As a result, wind energy usage has experienced a remarkably dynamic boom. Between 2002 and 2006, the installed power has more than doubled to approximately 48 000 MW. Two thirds of the European wind power are installed in Germany and Spain. And considering the world market, the USA represents the third largest market for wind energy plants (15.6% global market share), behind Germany (28%) and Spain (15.6%).

Considerable growth on the North American continent can be expected also for the coming years because the wind energy boom in the USA is continuing. The newly installed electric energy production capacity is expected to increase to more than 3000 MW in 2007, the American Wind Energy Association (AWEA) says. In the meantime, this is already leading to shortages because the number of suppliers is insufficient to cover the market's needs. This is an opportunity particularly for the European wind energy plant manufacturers and suppliers to foster exports to the United States. In turn, the European manufacturers can help US-American energy producers to overcome these bottlenecks.

The leading wind energy plant manufacturers include Danish and German companies. Particularly the German example shows how massive support of renewable energy promoted the rise of completely new high-tech industrial sectors that have grown to economically successful global players. This is why we will take a look at the interesting example of the German development in more detail.

Successful German Policy

In Germany, renewable energy (RE) sources have rapidly increased in importance in recent years, especially for the production of electric power [1]. In the year 2005, 10.2 % of the power from German electrical sockets originated from renewable sources, nearly three times more even than in 1990 [2]. This is due for the most part to the increased use of wind power. With an energy output of about 26.5 TWh per year, it has surpassed traditional hydroelectric power at 21.5 TWh (Figure 1). Power production using the biomass has also reached an all-time high of 13.1 TWh, and with an increasing tendency. Photovoltaic and geothermal power generation still play only a marginal role; nevertheless, power production by photovoltaic cells is growing rapidly: in Germany since the year 2000, it has increased by a factor of 15.

Germany is thus well on the way towards attaining the goal set by the previous Federal Government, that by the year 2010, renewable energy sources should supply at least 12.5 % of the power demands. The more distant goal of at least 20 % by 2020 could, with the current rate of growth, even be exceeded.

Heating from renewable energy sources is also on the rise. In 2005, the proportion of space heating from renewable sources in Germany was 5.4 %. The biomass is the unchallenged leader, with annually almost 76.5 TWh. Here, traditional wood burning is complemented by modern methods, for example wood-pellet heating systems. Solar thermal heating using collectors on roofs and other surfaces

INTERNET

Brochures [1,2] and other materials of the German Government (in English)
www.erneuerbare-energien.de

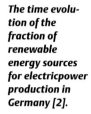

The time evolution of the fraction of renewable energy sources for electricpower production in Germany [2].

FIG. 1 | POWER FROM RENEWABLE ENERGY SOURCES

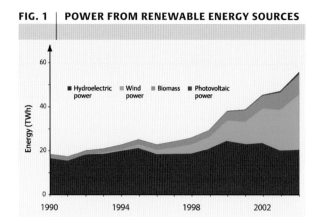

is also growing especially rapidly. Since 2000, it has more than doubled, and since the beginning of the 1990's, the increase has been more than twentyfold.

In the transportation sector, also, renewable energy carriers are slowly gaining ground. Automobile engines in Germany to be sure still consume a comparatively small fraction of biogenic fuels, mainly biodiesel, and small amounts of bioethanol. The proportion on the roads is only 3.4 % of the overall fuel consumption, but that is nevertheless a factor of eight more than in the year 2000.

The Current Situation

Where do the renewable energy sources stand overall in Germany? Figure 2 shows the distribution of the primary energy consumption in Germany in the year 2005. It should not be surprising that fossil fuels still dominate the energy supply: brown coal, anthracite coal, petroleum and natural gas meet 82 % of the requirements. In addition, nuclear power delivers 12 % of the electric power generated. The renewable energy sources supplied only 3.6 % of the primary energy consumption; in the mid 1990's, this figure was only about 1.5 %. More than half of this renewable energy is due to the use of the biomass; wind energy contributes 14.6 % and water power 11.9 % (Figure 2).

The reason for the strong increase in the use of renewable energy sources in Germany is to be found mainly in political decisions. In the past ten years, a public legal and

economic framework was set up to give the renewable energies a chance to gain a foothold in the market, in spite of their still relatively high power-generating costs. Along with various subsidy programs and the Federal Market Launch Program, these included in particular the Act on the Sale of Electricity to the Grid (StrEG) in 1991 and the Renewable Energies Law (EEG) in 2000, which set this development in motion. The principle: Power generated from renewable sources can be sold to the public power grid preferentially, and will receive a guaranteed price. The costs are covered by adjusting the price of the power sold.

The prices paid for renewable-source power are scaled according to the source and other particular requirements of the individual energy carriers. They are graded on a declining scale, i.e. they decrease from year to year. This is intended to force the renewable energy technologies to reduce their costs and to become competitive on the energy market in the medium term. The renewable energy technologies can accomplish this only through temporary subsidies, such as were given in the past to other energy technologies like nuclear energy. The renewable energy technologies will become strong pillars of the energy supply in the course of the 21st century only if they can demonstrate that they operate reliably in practice and are economically viable. To this end, each technology must go down the long road of research and development, past the pilot and demonstration plant stages, and finally become competitive on the energy market. This process requires public subsidies.

Potential and Limits

Often, the potential of the various technologies which exploit renewable energy sources is regarded with skepticism. Can renewable energies really make a decisive contribution towards satiating the increasing worldwide appetite for energy? Are there not physical, technical, ecological and infrastructural barriers to their use?

Fundamentally, their potential is enormous. Most of the renewable energies are fed by solar sources, and the Sun supplies a continuous energy flux of over 1.3 kW/m^2 (0.12 kW/ft^2) at the surface of the Earth. Geothermal energy makes use of the heat from within the Earth, which is fed mainly by radioactive decay processes (see p. 57).

These energy sources are, to be sure, far from being readily usable. Conversion processes, limited efficiencies, and the required size of installations give rise to technical restrictions. In addition, there are limits due to the infrastructure, for example the local character of geothemal sources, limited transport radius for biogenic fuels, the availability of land and competition for its use. Not least, the limited availability and reliability of the energy supplies from fluctuating sources play a significant role. Furthermore, renewable energies should be ecologically compatible. Their requirements for land, potential damage to water sources and the protection of the landscape and the oceans set additional limits. All this means that the natural, global supply

FIG. 2 | PRIMARY ENERGY USE AND SUPPLY IN GERMANY

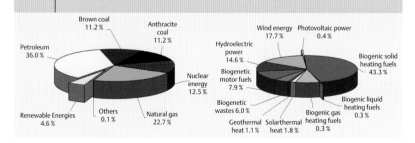

Left: The distribution among the various energy carriers to primary energy consumption in Germany; all together, 14 238 PJ (Petajoule) was used in the year 2005. Right: The contribution from different renewable energy sources in 2005; all together they produced 164 TWh of heat and electrical energy [2].

of potential renewable energies and the technically feasible energy production from each source lie far apart (Figure 3).

In spite of these limitations, a widespread supply of renewable energies is possible. In order for it to be reliable and stable, it must be composed of the broadest possible mixture of different renewable energy sources. In principle, water and wind power, use of the biomass, solar energy and geothermal heat can together supply all the requirements. Germany is a good example of this. Although it is not located in the sunny South, and has only limited resources in the areas of hydroelectric and geothermal power, nevertheless renewable energy sources can supply a considerable portion of Germany's energy requirements. Estimates put this contribution at up to 6200 PJ per year [1]. This corresponds to about 40 % of the current primary energy consumption. In these estimates, the boundary conditions in the form of usable land areas for collectors and solar cells, for wind parks and the cultivation of energy-yield plants (biomass) were set very conservatively.

Taking into account that in the coming decades, many technologies will become much more efficient in terms of their energy consumption, Germany could potentially supply 60 % of its energy needs from renewable sources within its own borders. The required broad and multiple uses of the renewable energy sources however also demands that the different sources be exploited according to their particular properties and limitations. Let us therefore take a closer look at the various types of renewable energy.

Water Power

Water is historically one of the oldest energy sources. Today, hydroelectric power in Germany comprises only a small contribution, which has remained stable for decades: 3 to 4 % of the electric power comes from storage and flowing water power plants. Its potential is rather limited in Germany, in contrast to the countries in the Alps such as Austria and Switzerland. In future, it will therefore be possible to develop it further to only a limited extent. Presently, the roughly 5500 large and small plants deliver about 25 TWh of energy annually; 90 % of this in Bavaria and Baden-Württemberg. The worldwide potential for hydroelectric power is considerably greater: nearly 18 % of the power generated comes from hydroelectric plants (see pp. 22). Thus, water power – considered globally– is at about the same level as nuclear power. So far, it is the only renewable energy source which contributes on a large scale to the world's requirements for electrical energy. In particular, "large-scale water power" is significant. An example is the Chinese Three Gorge Project, which will generate more than 18 GW of electric power, corresponding to about 14 nuclear power plant blocks (see pp. 22).

In Germany the so-called "small-scale" water power still has limited possibilities for further development. New construction and modernization of this type of water power plants with output power under 1 MW has ecological limits, since it makes use of small rivers and streams and it can affect the ecosystems. This alone strictly limits the possibilities for further development.

The advantages of water power are obvious: The energy is usually available all the time, and water power plants have very long operating lifetimes. Furthermore, water turbines are extremely efficient, and can convert up to 90 % of the kinetic energy of the flowing water into electric power (see pp. 22). By comparison: modern natural gas combi-power plants have efficiencies of 60 %, and light-water reactors have only about 33 % efficiency.

Wind Energy

The second important renewable energy source is the wind. Modern wind power plants, whose rotors operate on the aerodynamic principle, achieve efficiencies of up to 50 %. Germany is world champion in the use of wind power: 17,574 wind power installations produced nearly 4.3 % of the electric power used here in 2005. Worldwide, nearly 48 GW of electrical wind power generating capacity are installed, more than a third of it in Germany. The rapid increase of recent years is slowing down, however, since the majority of suitable sites on land have already been developed. The next step will therefore be the construction of offshore wind energy installations on the oceans.

Wind energy has been criticized in particular because of noise pollution, disturbance of animal life, especially the birds, and blighting of the landscape. Furthermore, wind has the disadvantage that it is not continuously available. This disadvantage can however be minimized by improved wind forecasts and intelligent input management into a decentrally organized power grid.

Biomass

The utilization of energy from the biomass is often underestimated. At present, biogenic heating fuels are being rediscovered in Germany. Wood, biowastes, liquid manure and other materials originating from plants and animals can be

FIG. 3 | NATURAL SUPPLY AND AVAILABILITY

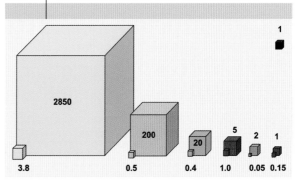

The natural supply of renewable energies in relation to the current world energy consumption (black cube, normalized to 1). Small cubes: The fraction of each energy source that is technically, economically and ecologically exploitable.
Yellow: solar radiation onto the continents; blue: wind; green: biomass; red: geothermal heat; dark blue: water power
(Source: DLR).

used for heating and also for electric power generation. The combination of the two uses is particularly efficient. In Germany, currently 94 % of the renewable heat originates from biofuels, mainly from wood burning – but increasingly also from wood waste, wood-chip and pellet heating and biogas plants. The contribution to power generation is also gradually increasing: in 2005, it was 2.1 %.

Biofuels are available around the clock and can be utilized in power plants like any other fuel. Biogenic motor fuels, as mentioned above, are getting renewable energy carriers rolling as suppliers for transportation.

Solar Energy

Solar energy is the renewable energy par excellence. Its simplest form is the use of solar heat from collectors, increasingly employed for household warm water heating and for public rooms such as sports halls and swimming pools. More than 7.2 million square meters of collectors are currently installed on German rooftops.

Solar-thermal power generation is, in contrast, still in the development phase (see pp. 26). Parabolic trough collectors, solar towers or paraboloid reflector installations can produce temperatures of over 1,000 °C (1,832 F), which with the aid of gas or steam turbines can be converted into electric power. These technologies could in the medium term contribute to the electric power supply. They are however efficient only in locations with a high level of insolation, such as Spain. Germany would thus have to import solar power from solar-thermal plants via the common power grid, which initially could be laid out on a European basis; in the long term, North African countries could supply solar power via a ring line around the Mediterranean Sea [1].

The most immediate and technologically attractive use of solar energy is certainly photovoltaic conversion. From the point of the energy economy, its contribution is still marginal – mainly due to the still high power generating costs. Nevertheless, the market for photovoltaic devices shows by far the most dynamic growth: In the year 2005 alone, a peak power capacity of 600 MWp was installed. All together, currently over 1,400 MWp generating capacity is installed in Germany. New production techniques at the same time offer the chance to produce solar cells considerably more cheaply and with less energy investment, and thus to allow a breakthrough onto the market (see pp. 42 and pp. 50).

Geothermal Energy

The renewable energy resource which at present is the least developed is geothermal heat. Deep-well geothermal energy makes use of either hot water from the depths of the Earth, or it utilizes hydraulic stimulation to inject water into hot, dry rock strata (hot-dry rock process) (see pp. 54), with wells of up to 5 km (3 mi) deep. At temperatures over 120 °C (248 F), electric power can also be produced – in Germany thus far only at the Neustadt-Glewe site in Mecklenburg-Vorpommern. Favorable regions with high thermal gradients are in particular the North German Plain, the North Alpine Molasse Basin, and the Upper Rhine Trench.

Geothermal heat has the advantage that it is available around the clock. However, the use of geothermal heat and power production is still in its infancy. Especially the exploitation of deep-well geothermal energy is technically challenging and still requires intense research and development (see pp. 54). If geothermal sources can be successfully utilized, then they could make a considerable contribution towards meeting the base demand in view of their uniformity and reliability.

The Window of Opportunity

How will energy supplies in Germany develop in the future? Will all the renewable energy source options play a role, and if so, to what extent? A glance at the current situation shows that the energy economy and in particular the electric power suppliers are facing important investment decisions, since the German power generating plant is getting old. By 2020, production capacity of 40 to 45 GW_{el} will have to be modernized or replaced. This is after all a third of today's total installed capacity.

Thus, at present a window of opportunity for investment in the construction of renewable energy plants is opening. This is independent of the decision concerning the extension of the remaining operating lives of some of the German nuclear power plants, since that would delay the necessary renewal by only a few years. After 2020 at the latest, Germany will have to be supplied by a broad energy mix from a variety of sources. Among these will be highly efficient fossil fuel power plants. According to the opinions of some energy experts, "CO$_2$-free" plants can likewise make a contribution; in them, the CO_2 will be separated out of the exhaust gases and stored.

How well the renewable energies will be able to establish themselves has been the subject of investigation of various studies and simulations. The result is that they will successively increase their contributions to the overall energy supply. As a rule, they are initially immature and cost-intensive and thus remain at a low level. Following technological and economic development, they then move in-

The development of land-besed wind power in Germany exhibits a nearly exponential growth during some periods of time. Recently, however, there are indications of a turning point towards lower growth rates (Source: Deutsches Windenergie-Institut).

FIG. 4 | WIND POWER INSTALLATIONS IN GERMANY

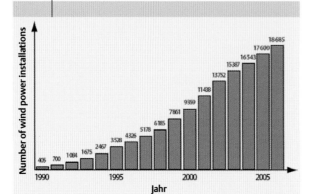

FIG. 5 | FUTURE POWER GENERATION

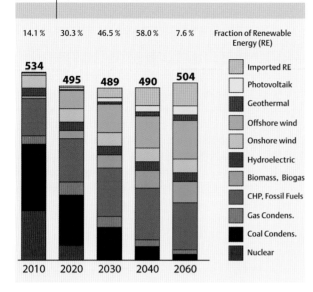

Electric power production in Germany according to type of power plant and energy source in the future scenario „Naturschutz Plus I" [9] (RE: Renewable Energies).

FIG. 6 | WORLD CONSUMPTION OF PRIMARY ENERGY

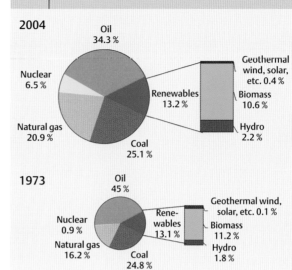

The evolution of primary energy consumption by the world's population: In 1973, the total consumption was 253 000 PJ, and in 2004, it was 463 000 PJ [4].

to a phase of steady, at times rapid and nearly exponential growth. Only when technological, economic or structural limiting factors come into play does the growth level off, and stabilizes in the long term at a constant level. A good example for this model is the development of land-based wind power. Figure 4 illustrates its rapid growth in the past several years in Germany. But the turning point has visibly been reached, since the economically attractive sites with strong winds on land have for the most part already been exploited. There will certainly be some continued construction of wind plants on land, but within a few years the evolution of the proportion of wind energy in the overall mix will have reached its saturation level. There can then still be some continued growth through "repowering", i.e. the modernisation of older installations with new, more efficient rotors and generators. This will lead to some further increases in the power production capacity of land-based wind power.

The High Sea and the Open Field

The next major step in the modification of the energy systems in Germany will most likely be the introduction of offshore wind energy. Along the German seacoasts and within the "exclusive economic zone", a potential power-generating capacity of up to 25 GW of electric power output is predicted.

Such offshore wind installations will have to be built far from the coastline in water depths of up to 60 m (200 ft). This is particularly true of the North Sea, which has strong winds. In the shallow water near the coasts, there are no suitable sites due to nature protection areas, traditional exploitation rights such as gravel production, restricted mili-

tary zones and ship traffic. Plants in deeper water, however, require a more complex technology and are more expensive. The high power sea cables for transporting the power to the coast over distances of 30 to 40 km (19-25 mi) will also drive up the investment costs.

Here, the offshore installations near the coast have a considerable advantage: the wind is stronger and steadier over the free water surface. This compensates to some extent for the higher costs of these wind parks. However, the individual plants must deliver high power outputs. Only when they produce around 5 MW$_{el}$ can they be economically operated under such conditions. To construct the first German offshore test field in the North Sea, industry, associations and power grid operators have jointly formed an Offshore Foundation.

This first offshore windpark is to be built in 2007 near the research platform 'Fino' in the neighborhood of Borkum West. If it operates successfully, wind parks can grow up out of the waves one after the other.

The biomass will play an increasingly important role in the medium term. Taking the requirements of nature and landscape protection into account, and with the expectation of ecological compatibility, then by 2020 the consequent utilization of biogenic wastes will be a main focus. This includes the thermal utilization of wood waste products and agricultural wastes as well as an increased deployment of biogas installations. An increasing and ecologically compatible utilization of the biomass will also have structural consequences. It opens up new, long-term perspectives and possibilities for agriculture and forest management. Many a farm landlord will thus become an "energy landlord".

FIG. 7 | WORLDWIDE USE OF RENEWABLE ENERGIES

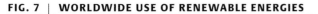

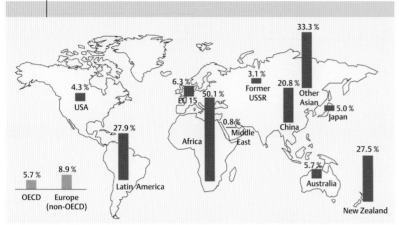

The fraction of renewable energies in the supply of primary energy to various regions in the year 2001 [4].

Liquid motor fuels from the biomass can substitute fossil fuels to a much greater extent than is presently the case in Germany. However, in the near term, there will be no new large agricultural areas for growing energy-yield plants. Protection of the fields from erosion, reservation of land for the interlinked biotopes which are legally required by the nature protection laws, and other sustainability goals will to a large extent prevent such land usage. After 2020, the population reduction and accompanying reduction in food requirements, as well as agricultural yield improvements, will probably allow significant increases in the production of energy-yield plants. This will in turn permit an increase in the proportion of biogenic motor fuels for transportation.

The remaining renewable energy carriers will be able to play a significant role in the energy economy in Germany only in the distant future. Photovoltaic power and imported power from solar-thermal power plants belong in this group.

Geothermal heat is currently a particularly hopeful energy source. It is, however, still too early to predict its future contribution to the primary energy supply, because of the technical challenges involved in its exploitation. The utilization of geothermal heat will become really efficient only when it is used for space heating in combination with the production of electric power (CHP, Combined Heat and Power). This will require the construction of local and regional pipe networks, i.e. additional investments.

Ecologically Optimized Development

Just how the proportion of renewable energies within the energy mix in Germany will evolve in reality cannot of course be precisely predicted. However, model calculations make it clear which paths this evolution might take under plausible assumptions. The Institute for Technical Thermodynamics at the DLR in Stuttgart carried out a comprehensive study in 2004, analyzing various scenarios [3]. They took technical developments, economic feasibility, supply security and eco-logical and social compatibility into account. This study shows up the essential trends.

Figure 5 gives the distribution of power generation in Germany according to the type of power plant and the energy source within the scenario "Naturschutz Plus I" [3]. This scenario aimed at an economically acceptable increase in the use of renewable energies, but also took ecological factors into account. Furthermore, it assumed that the utilisation of nuclear energy will come to an end in Germany between 2020 and 2030.

The developers of this scenario came to the conclusion that electric power generation, which at present is supplied mainly by fossil fuels and nuclear reactors, will be renewed step by step with sustainable energies.

In this scenario, coal-fired power plants, gas power plants and especially combined heat and power generation (CHP) on the basis of fossil fuels will play an important role far into the 21st century. Towards the middle of the century at the latest, the proportion of electric power from renewable energy sources should surpass the fifty percent level. Major contributions will be made by offshore wind parks and power generation from the biomass and biogas. The contributions of photovoltaics and geothermal heat will be perceptible but not of primary importance. In addition, imported power from renewable sources will be significant.

It is however also clear that the changes in the power mix will have to be accompanied by a clearcut improvement in overall energy efficiency and thus a decrease in the total power consumption. Only when energy conservation, improved efficiency, and the increased utilization of renewable energies are all promoted at the same time will a sustainable energy supply for Germany become feasible.

Renewable Energies on a Worldwide Scale

A long-term solution in terms of climate policy, supply security and an equitable access to energy can be achieved only on a global scale. Today, we are far from such a solution (Figure 6). The proportion of renewable energies in the world's primary energy consumption is currently 13.4 %. It has remained practically constant since the beginning of the 1970's. This is due to the increased consumption by the world's population, since the overall capacity from renewable energy carriers has nearly doubled in the same period.

The utilization of renewable energies varies widely in different regions. In some European countries and in North and South America and Japan, conventional hydroelectric power traditionally plays a strong role. Countries such as Austria, Switzerland, Norway or Canada profit from their favorable topographical situations. Hydroelectric power dominates the contribution of renewable energies to the worldwide electric power supply.

Worldwide, approximately 60% of the renewable energy is consumed for heating in private homes, the public sector, as well as the service sector. The use of wood and

FIG. 8 | WORLDWIDE USE OF RENEWABLE ENERGIES 2004

Transition countries: countries in the transition from state-directed economy to market economy; under this category, the IEA summarizes countries from non-OECD Europe and the countries of the former USSR.

1) Biogenic portion of waste; in the non-OECD countries, a clear distinction between biogenic and non-biogenic waste is not always possible.

2) Geothermal, solar, wind, ocean.

3) Latin America without Mexico and Asia without China.

	PEC	RE thereof	Share RE of PEV	Shares of most important RF of total RE [%]		
	[PJ]	[PJ]	[%]	Hydro	Biomasse / waste 1)	Others 2)
Africa	24,535	12,021	49.0	2.6	97.0	0.4
Latin America 3)	20,327	5,870	28.9	36.1	62.4	1.4
Asia 3)	53,986	17,187	31.8	4.0	92.4	3.6
China	68,100	10,509	15.4	12.1	87.9	0.0
Middle East	20,089	138	0.7	43.4	32.2	24.4
Transition countries	45,369	1,712	3.8	63.7	34.6	1.6
OECD	230,610	13,189	5.7	34.6	53.4	12.0
World	**463,017**	**60,626**	**13.1**	**16.7**	**79.4**	**4.0**

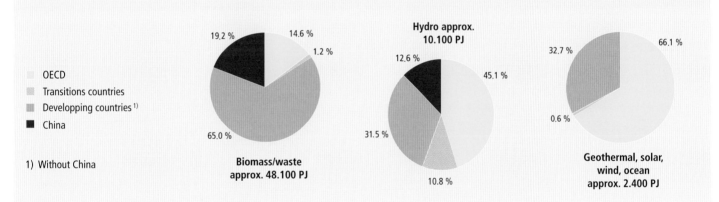

OECD
Transitions countries
Developping countries 1)
China

1) Without China

Global overview of regional usage of renewable energies (Source: IEA).

charcoal is dominant. Whereas the Western industrial countries (OECD) use half of the renewable energy sources for production of electricity, the non-OECD countries use only 14.1% for this purpose. Here, the share of approximately 70% used for decentralized heat supply is correspondingly high, as opposed to about 18% in the OECD countries.

In the developing countries, renewable energy sources do not have exclusively positive effects. This is shown in part by the history of large dams, with their often very negative social consequences and their impact on local ecosystems. A second serious problem is the traditional burning of biomass. It makes the statistical contribution of renewable sources to the primary energy supply appear high, especially in Africa, but also in many countries in Asia (Figure 8). However, it is not really sustainable, since the forests are often irreversibly cleared in the process. It is thus especially

important for developing and emerging nations that renewable energy sources be utilized with a critical view to the local conditions.

Investment Costs

The discussion of the use of renewable energy sources takes place within a balancing act between technical arguments on the one hand, and economic and political arguments on the other. This becomes particularly clear when one considers the question of costs.

Most renewable energy sources today are more expensive than the corresponding fossil-fuel and nuclear sources. That is, as described above, the reason for the financial subsidies to renewable energies, be it through pay-as-you-go financing as in the case of electric power, or be it through targeted subsidies as in the market launch program of the

German Federal government for the area of renewable heating sources. The goal remains to provide an incentive for the reduction of the specific costs; therefore the degressive price policy in all cases. However, the amount of these subsidies is often overestimated. This is particularly true of the electric power sector, which in Germany is characterized by massive economic interests and an oligopoly market structure. A detailed treatment of this subject can be found in the publication of the German Ministry for the Environment, Natural Conservation and Nuclear Safety (BMU) [5].

According to a study made by the Institute of Physics at the University of Oldenburg, the investment costs for wind energy installations in Germany are on the average about 1000 €/kW (approx. 1,400 $/kW). For coal-fired plants, they are presently 800 to 1000 €/kW (1,200-1,400 $/kW), and for the nuclear power plant currently being planned in Finland, they are estimated to be 1700 to 2000 €/kW (2,500 - 2,900 $/kW). Here, external consequential costs are not included: Additional facilities for minimizing the emissions of CO_2, for example by separation and storage of carbon, would make fossil-fuel power plants considerably more expensive. Economically more relevant are, in the end, the costs per kilowatt hour generated. Here, for the conventional energy sources, the costs of the fuel must be added to the investment costs; the former are not relevant for wind power, hydroelectric power, or solar energy.

The Overall Energy Balance

The overall energy balance for construction, operation and disposal of the plants is in fact very positive for all sources of renewable energy. Calculations by the DLR in Stuttgart and the ifeu-Institute in Heidelberg showed energy amortisation times of 3 to 7 months for wind energy plants, 9 to 13 months for hydroelectric plants, 3 to 7 months for solar-thermal power plants, and 1.5 to 2.5 years for solar collectors. Most energy-costly are currently photovoltaic installations; they are amortized in terms of energy only after 2 to 3 years (thin-film cells) and 3 to 5 years (polycrystalline silicon) of operation. The lifespan of all these installations is however longer by a large factor. Power plants and boilers based on non-renewable energy sources are, by the way, never amortized in terms of energy, since they always consume more fuel energy than they produce as useful energy.

It is true that the storage of renewably-produced energy is an important challenge. Research is being carried out in Germany on a number of technologies for this purpose. Whether or not the use of renewably-produced hydrogen will play a decisive role for this storage is not yet clear. It is however certain that these potential methods will remain economically marginal in the coming one or two decades. Much more suitable and cost-effective at present is intelligent management on the supply and the user sides, and of course the application of the many still untapped possibilities for increasing the efficiency of energy conversion and energy use.

The evolution of the energy prices, the increasingly obvious interdependence of energy policy and world politics, and the resulting risks, are all signs that the importance of renewable sources for our energy supply could increase much more rapidly than thus far expected.

Technology for the 21st Century

Nevertheless, renewable energy sources will dominate the energy technology of the 21st century. Even in Central Europe, which is not blessed by the Sun, there is all together an enormous potential for renewable energies at our disposal. This will make them in the medium and long term the winners by points relative to fossil-fuel and nuclear technologies in decisive areas which range from geopolitically sensitive dependencies to economic feasibility. Conventional, in particular fossil-fuel energy sources such as petroleum will become much more costly, while the climate- and resource-political advantages of the renewable energy sources will gradually make themselves felt also in their prices. Then, at the latest, their ascendancy will be unstoppable.

Summary

In the year 2005, 4.6% of the total primary energy consumed in Germany was supplied from renewable energy sources; for electrical energy, the proportion was 10.2%. The main reason for this is the boom in wind power, which will continue to expand, especially due to offshore wind energy parks in the oceans. Hydroelectric power has traditionally provided a significant contribution to electric power generation, but its potential for expansion is limited. Photovoltaics, solar-thermal and geothermal power generation at present play only a minor role. In Germany in 2004, 5.4% of the thermal energy requirements were filled from renewable sources, for the most part from the biomass. Heat production from solar-thermal systems has more than doubled since the year 2000. For vehicular transport, biofuels play a minor role at 5.4%. By 2050, the proportion of primary energy obtained from renewable sources could reach 50% in Germany.

References

[1] Brochure of the German Ministry for the Environment, Natural Conservation and Nuclear Safety (BMU): Renewable Energies – Innovation for the future. BMU, Berlin 2006. PDF download available at www.erneuerbare-energien.de/inhalt/37453/20049

[2] Brochure of the German Ministry for the Environment, Natural Conservation and Nuclear Safety (BMU): Renewable energy sources in figures – national and international development. BMU, Berlin June 2007. PDF download available at www.erneuerbare-energien.de/inhalt/5996

[3] J. Nitsch et al., Ecologically Optimized Extension of Renewable Energy Utilization in Germany. German Aerospace Center (DLR), Stuttgart 2004. A short and the long version are available at www.dlr.de/tt/en/desktopdefault.aspx/tabid-290

[4] International Energy Agency, Renewables Information Outlook 2006. IEA/OECD, Paris 2006.

[5] Brochure of the German Ministry for the Environment, Natural Conservation and Nuclear Safety (BMU): What electricity from renewable energies costs. BMU, Berlin 2006. PDF download available at www.erneuerbare-energien.de/inhalt/36865/20049

The publications of the BMU can be ordered from the Public Relations Department in Berlin or at www.erneuerbare-energien.de.

About the Author

Harald Kohl, born in 1963, studied physics in Heidelberg and carried out his doctoral work at the Max-Planck Institute for Nuclear Physics there. Since 1992, he has been a technical officer in the Federal Ministry for the Environment, Natural Conservation and Nuclear Safety (BMU) in Bonn and Berlin. He is currently responsible for fundamental aspects of renewable energy sources.

Contact:
Dr. Harald Kohl,
Bundesministerium für Umwelt, Naturschutz und Reaktorsicherheit,
Alexanderplatz 6, D-10178 Berlin
Germany.
harald.kohl@bmu.bund.de

Wind Energy

A Tailwind for Sustainable Technology

BY MARTIN KÜHN

The use of wind energy has experienced a rapid increase in the past fifteen years. What technologies, economic and political factors have fostered this development? Will further increases be compatible with the present system of energy supplies?

By mid-year of 2007, there were more than 19 000 wind turbines with a total capacity of nearly 21 GW installed in Germany – and the trend is for this number to increase. Assuming an average wind year, these installations together can generate 6.9 % of the total net electricity consumption, and thus exceed the power generated in Germany today from every other renewable energy form. More than a quarter of the worldwide wind power is at present installed in Germany, but the international markets are also growing. Within the last five years, the annual newly-installed capacity increased by 17 % on average, especially in other European countries, the USA and in Asia (Figure 1). The German wind energy sector currently provides 74 000 jobs and exports 74 % of the turbine equipment produced; it is profiting effectively from the continuing international boom.

From the Drag Device to the High Tip Speed Turbine

Humans have made use of wind power for around 4000 years. Besides sailing ships, wind-driven pumps and mills were developed long ago. Early forms of windmills used a rotor with a vertical axis which was driven by the drag force exerted by the air flow passing the rotor blades. This design concept, known as a drag device, has a low efficiency, at most about one fourth of that of the lift devices

INTERNET

Federal Ministry for the Environment
www.erneuerbare-energien.de

European Wind Energy Association
www.ewea.org

Endowed Chair for Wind Energy at the University of Stuttgart
www.uni-stuttgart.de/windenergie

FIG. 1 | DEVELOPMENT

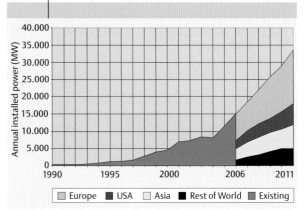

Worldwide time development of the annual newlyinstalled power from wind energy since 1990, and the prognosis up to 2010 *(Graphics: BTM Consult ApS).*

described in the following [1]. Today, it is therefore used practically only in the form of the cup anemometers that measure wind speed. From around the 12th century on, new windmill types were developed, such as the post mill or Dutch windmill, which operated on the basis of a different and more effective principle. The decisive advance was not the generally horizontal orientation of the rotor axis, but rather the fact that the flowing air drives the rotor blades via the aerodynamic lift force. For a drag device, which is moving with the flow, the relative velocity at the rotor blades is always smaller than the wind speed. Lift devices, in contrast, can achieve higher apparent wind speeds by superposition of the wind speed and the circumferential velocity of the rotor. Only in this way can the forces necessary for an optimal deceleration of the wind be generated, and the aerodynamic efficiency approaches its theoretical maximum of 59 % calculated by Albert Betz and F.W. Lancaster.

For the generation of electric power, along with an increase in the aerodynamic efficiency, it is also a constructional advantage if the circumferential rotor velocity is much greater than the wind velocity. In these so-called high tip speed turbines, only a low number of very slim rotor blades are necessary, and the generator can be driven at high rpm with a correspondingly low torque. The use of three rotor

blades has become standard for structural dynamical, acoustic, and aesthetic reasons.

From Grid-connected Wind Turbines to Grid-supporting Wind Power Plants

Even though the external appearance of wind turbines has not changed much in the past 15 to 20 years, inside their nacelle enclosures there has been a rapid technical development. Increasingly larger and more efficient wind turbines are feeding electrical energy of improved quality and at lower cost into the power grid. This becomes clear if we consider the various design concepts currently in use. Every installation requires a method of limiting the power and the load on the turbine, since the power in the wind increases as the third power of its velocity. Two principles have established themselves for providing this control mechanism: stall and pitch.

In the simplest design, the rotor blades are attached to the hub in a fixed position (Figure 2). The speed of rotation is held nearly constant by an induction generator coupled directly to the power grid. This is typically a three-phase asynchronous motor operated in generator mode. When the wind gets stronger, its angle of attack at the rotor blades changes as a result of the vector addition of the wind velocity and the circumferential velocity of the rotor. This increase in the angle of attack leads to a flow separation on the low-pressure side of the blades, and thus to stall. If the blades were the wings of an aircraft, it would be in

danger of crashing. The wind turbine, in contrast, is protected in this way from excessive power, since the driving lift forces acting on the blades are reduced and the drag is increased (Figure 3). This so-called "Danish" concept was introduced in 1957 by the Danish wind power pioneer Johannes Juul. Due to its simplicity and robustness, it was important in the first deployment of wind turbines in large numbers in the mid 1980's, with rotor diameters of 15 to

FIG. 2 | A STALL-REGULATED WIND TURBINE

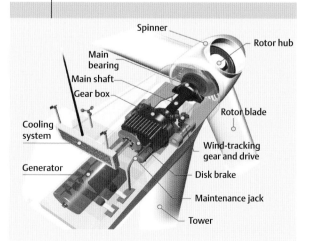

Machine layout of a stall-regulated wind turbine with gear box and constant rotation speed, designed by NEG-Micon (Graphics: German Wind Energy Association).

FIG. 3 | THE STALL CONCEPT

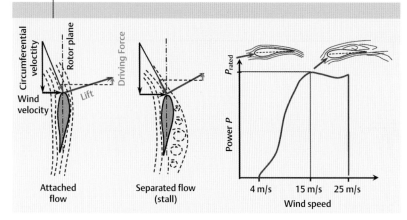

Left: Power limitation with increasing wind speed by flow separation or stall; right: Power curve, showing the limitation of power above rated wind speed.

20 m. About ten years later, the concept was developed further by mounting the rotor blades on a bearing where they are attached to the hub. The angle of attack of the apparent wind speed is increased by only a few degrees through turning the trailing edge of the blades into the wind direction. So, the separation can be actively controlled and the desired maximum power can be reliably regulated (active-stall concept).

The second principle for limiting power is based on a greater variation of the rotor blade angle, or pitch angle. If the wind speed increases after the rated power has been reached, then the leading edge of the rotor blades is turned into the wind (Figure 4). By decreasing the angle of attack, the power and the load are limited. This concept, oriented towards lightweight design, was decisively influenced by the wind energy pioneer Ulrich Hütter from the University of Stuttgart, who in 1957 constructed a pitch-regulated two-bladed turbine. At this trendsetter for the first time, rotor blades made of fibreglass reinforced plastic were used.

FIG. 4 | THE PITCH CONCEPT

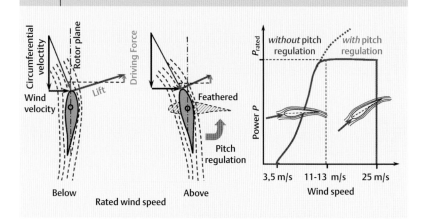

Left: Power control using pitch regulation; right: Power curve.

It soon became clear that turbines which operate at a constant rotational speed cannot completely compensate a gusty wind even if the rotor blades are adjusted very quickly, and that those machines were thus subject to strong short-term variations in power output, accompanied by structural stresses and corresponding reactions on the power grid. The advantages of the pitch concept, i.e. a constant rated power and favourable behaviour during startup and during storms, can be put into practice only in combination with certain variability in the rotational speed of the turbine. This however requires some additional effort in the design of the electrical system. To this end, initially three, and now two machine concepts have become common.

Initially, in particular the Danish firm Vestas introduced a concept that allows the variation of the rotational speed by up to ten percent. This is accomplished by a fast regulation of the rotational speed compliance (slip) of the asynchronous generator, which is coupled directly to the power grid. Through the interactions of the turbine rotor, which now acts as a large flywheel, with the somewhat slower pitch adjustment, wind variations above the rated operating velocity can be smoothed out quite satisfactorily.

Especially in Germany, beginning in the 1980's with experimental installations and commercially from 1995 on, a concept involving complete variability of the rotational speed was developed, which today is used in more than half of all new plants. While the stator of the asynchronous generator is still coupled directly to the power grid, the rotor accepts or induces precisely the AC frequency which is required to adapt to the desired rotational speed. By means of such a double-feed asynchronous generator, the rotational speed can be roughly doubled between the startup velocity of about 3.5 m/s and the rated operating velocity of 11 to 13 m/s. In the wind speed range the rotor operates near its aerodynamical optimum, and aerodynamic noise is effectively reduced. At full load conditions, the rotational speed oscillates by approx. ±10 %, in order to smooth out wind gusts, again in combination with the pitch control.

The most evident, but complex path to complete variability of the rotational speed lies in electrically decoupling the generator from the grid using a frequency converter with an intermediate DC circuit. In this concept, in which as a rule a synchronous generator is employed, all of the electrical power is passed through the frequency converter. By controlling the excitation in the generator rotor, the rotational speed can be varied up to three times its startup value.

The Enercon company, market leader in Germany, applies this concept very successfully to installations without a mechanical transmission using a specially developed direct-drive multi-pole synchronous generator (Figure 5). In recent years, this principle, owing to its excellent grid compatibility and its independence from the local grid frequency, has also been applied to some geared machines, which still supply approx. 85 % of the world market.

FIG. 5 | PITCH-CONTROLLED WIND TURBINE

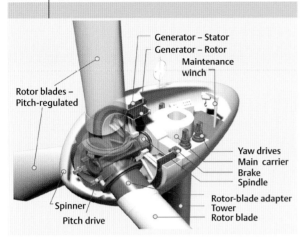

Generator – Stator
Generator – Rotor
Maintenance winch

Rotor blades –
Pitch-regulated

Yaw drives
Main carrier
Brake
Spindle

Spinner
Pitch drive

Rotor-blade adapter
Tower
Rotor blade

Machine layout of a variable-speed, pitch-controlled wind turbine without a gear box, built by the Enercon company (Graphics: German Wind Energy Association).

In the meantime, the latter two concepts of variable pitch and variable rotational speed have asserted themselves in the market and have practically superseded the simple, robust stall-regulated turbines of earlier days. Partial or complete decoupling of the generator from the power grid provides a great improvement in grid compatibility and even – under favourable circumstances – allows the support of the electrical power grid. The phase angle between the current and the voltage (power factor) can be adjusted as needed. Negative effects on the grid, such as switching currents, voltage and power variations and harmonics, can be avoided or strongly reduced. Furthermore, the machines are much less sensitive towards disturbances from the grid, such as short-term voltage breakdown.

Lightweight Design, Advanced Controls, and Reliability

Fifteen years ago, stall-regulated turbines were technically still for the most parts comparable to agricultural machines. Today's wind turbines, with rotor diameters of up to 126 m, are among the largest rotating and vibration-prone machines made. They defy the extremely harsh environmental conditions in the atmospheric boundary layer near the ground by employing complex automatic control systems, for example by monitoring a number of different operating pa0thermore, the most modern structural materials are used, such as carbon fibre composites or fatigue resistant cast iron and forged steel.

Due to the temporal and spatial structure of wind fields, every local gust has a multiple effect on the rotating blades. Within the planned lifetime of twenty years for a wind turbine, up to a thousand million load variations therefore occur – this is a completely unknown order of magnitude in other areas of mechanical engineering. At the same time, the increasing rotor size requires more lightweight design methods; otherwise, the materials stresses resulting from

the continual alternating bending moment from the self-weight of the rotor blades would become a problem.

In terms of commercial competitiveness compared to conventional power plants, the need for further cost savings also drives the developments. They can be achieved not only by economies of scale through production of large numbers, but also by increasing the efficiency of the individual turbines. Since generally the maximum theoretical aerodynamic efficiency is already approached quite closely, one tries to further reduce the investment costs per kilowatt hour generated. This can be achieved for example through active and passive vibration damping, mitigation of dynamic loads, and the application of lightweight concepts. In addition, the operating costs can be decreased by a further improvement of the reliability of the installations. The technical availability of installations, i.e. the proportion of the time during which the turbines are operable, lies in the meantime near 98-99 % [2]. Nevertheless, further improvements in the durability of expensive components like gear boxes and the reliability of sensors are necessary. This is particularly true for turbines in the megawatt class. Such machines have been installed in large numbers since the end of the 1990's and at the beginning of this decade, often after only an all too brief try-out period.

Offshore Wind Energy

In the near-coastal regions of the oceans, there are enormous wind resources waiting to be tapped. Besides a higher energy yield by 40 to 50 % as compared to good coastal sites, here a greater site area is available. The Federal Ministry for the Environment in Germany expects the installation of 15 GW offshore within the coming fifteen years, compared to 10 GW on land.

Following the first ideas for offshore wind projects in the 1970's, during the 1990's several smaller European demonstration projects were set up. After 2000, the construction of commercial wind farms with up to 160 MW power took place. By the end of 2006, the installed offshore capacity was over 900 MW. This corresponds to over 1.5 % of the worldwide installed rated power. Operating experience has so far been mainly positive and supports further development, which presently is taking place, in particular in Great Britain, Denmark, the Netherlands and Sweden. On the other hand, awareness of the challenges of this complex technology is also increasing. 2004, in the largest Danish offshore wind farm at Horns Rev, only two years after commissioning, all eighty nacelles had to be temporarily taken down and overhauled on land at considerable expense. This however also demonstrated that the industry is in the meantime sufficiently mature to survive impacts of this magnitude.

In Germany, the water depths of 25 to 40 m and the distances to shore of 30 to over 100 km in suitable areas represent especially a financial hurdle for the initial projects. In spite of the large number of proposed and to some extent already permitted offshore wind projects, the first ac-

Fig. 6 *Installation of a 5 MW offshore wind turbine, off the Scottish coast in August 2006. The rotor diameter is 126 m* (Photo: REpower System AG).

tual realisation at the offshore test field 45 km north of the island of Borkum will not happen before 2008. There, twelve turbines of the powerful 5MW class are planned; they are currently available from only three German manufacturers. In the year 2006, a turbine of this type was installed on a jacket support structure in a water depth of 44 m off the Scottish coast (Figure 6).

Grid Integration of Dispersed and Intermittent Wind Generation

In general, it is expected that a generation proportion of up to 20 % from renewable energy sources such as wind and solar power can be integrated into a power grid without major problems. Although this situation will occur in Germany on an annual and geographic average only after about ten years, the integration of new wind power plants is already today presenting technical and economic challenges. This is due to the regional concentration of wind turbines in the Northern and Eastern coastal areas, and to the daily and seasonal wind variations. From time to time the wind generation there exceeds the local grid demand, while at other times, there is almost no wind power available.

Dispersed generation, e.g. by wind turbines, into the weak periphery of the power grid, new power generating and power consuming facilities, and the liberalisation of the market require a reorganisation of the decades-old structure of the European electricity supply grid into a transport grid for large amounts of traded energy. A study carried out by the German Energy Agency (DENA) in cooperation with the energy producing sector and the wind energy industry has investigated the consequences of increasing the proportion of wind power to 15 % by the year 2015. According to this study, there are no serious technical problems, and only moderate additional costs are to be expected. About 400 km of the current 380kV grid will need to be upgraded, and around 850 km must be newly constructed. This corresponds to 5 % of the present power transmission

network. In ten years, the additional costs, depending on the scenario, will be between 0.39 and 0.49 ct/kWh for private households and 0.15 ct/kWh for the industry. This estimate contains the costs for upgrading the grid as well as balancing power, minus the avoided costs of conventional power generation [3].

Since the year 2003, for new installations in regions with a high wind energy penetration, a gernation management has been applied which permits the operators of the transmission system to reduce or switch off individual power producers when the grid load is too low or when grid shortfalls occur. For conventional power plants, this practice leads to a savings in the cost of fuel and operations. For wind power producers, in contrast, it can give rise to serious losses of revenues, since here, the operating and financing costs remain nearly constant.

New turbines also require additional capacity in the power transmission network. But the planning of new overhead lines is hampered by public acceptance problems and protracted authorisation procedures. New solutions, such as conventional buried cables or new bipolar AC cable concepts with a high transmission capacity, are being adopted only hesitantly by the transmission system operators. However, there are still considerable capacity reserves in the present supply grid, if the effective thermal transmission capacity is exploited in periods of cool weather or strong winds. The measurement of weather data could permit transmission of 30 % more power and with monitoring of the transmission line temperature, the increase could be up to 100 % [4]. In Germany, monitoring of this type is being introduced in Schleswig-Holstein for the first time, and it has been practiced in some other EU countries for several years.

The grid management of the power grid by the four German transmission system operators (TSOs) consists mainly of a continuous adaptation of the generated power supply to the varying demand. Power generation and purchases are planned 24 hours in advance. By controlling on and off of power plants with different response time constants, and by short-term buffering using the rotational energy of the generators and turbines, equilibrium is maintained. While up to now, only the load variations and possible power plant malfunctions had to be compensated, in the future the grid management will be complicated by the intermittent contribution from wind energy, which has a priority purchase and transmission status. Wind energy prediction models are employed in order to minimise the required capacity of conventional power plants and of additional balancing power. At present, the average deviation of the 24 hour predictions is about 6.5% (expressed as the mean square error normalised to the installed wind power capacity) [5].

Considerable deviations in the predictions can occur in particular due to time offsets in the passage of weather fronts and the corresponding significant power gradients. Under such unfavourable conditions, the wind generation within a regulation zone can decrease by up to 1 GW per hour and by several gigawatts within a few hours. Further improvements of the forecasts and a reduction of balancing power would be possible by using new communications technology, by introducing a more flexible power plant planning, and by short-term balancing among the different transmission system operators. Reasonable measures include a short-term correction of the 24 hour predictions, online measurements of the wind generation, and the introduction of shorter trading periods in the power market (intraday trading). The DENA grid study found that by the year 2015, no additional balancing power plants will be required. Furthermore, on the average an hourly and minute reserve of conventional power plant capacity amounting to 8 to 9 % of the installed wind energy capacity should suffice.

In order to maintain the traditionally very good network stability and supply security in Germany, new grid codes for wind energy generators were introduced in 2003; they require the turbines to meet certain criteria. Older, previously installed wind energy machines which correspond to the earlier criteria have to be shut down immediately if network malfunctions occur. This could, in unfavourable cases, lead to a sudden deficit of several gigawatts of input power and produce instabilities in the European electric power network. These risks can however be minimised by modern wind turbines with converter technology, by retrofitting of older installations, and by modernisation of the transmission grid, which is in any case necessary. Network stability and security can thus be guaranteed even with further increases in the proportion of wind energy.

An increasing proportion of wind generation, with its quasi day-to-day variability, will in the medium term require energy storage systems on the scale of power plants. The construction of new pump storage hydro power plants in Germany is not to be expected in the future. Storage via electrolytically produced hydrogen as an alternative has a very low system efficiency. In the foreseeable future, it will be more reasonable to save fossil fuel by making use of wind energy, and to tide over wind variations by using conventional power plants (see pp. 83). Underground adiabatic compressed air energy storage (CAES) systems with thermal energy retrieval, which can yield efficiencies of up to 70 %, have relatively good future prospects. However, the application of this completely new technology cannot be expected before the year 2015.

Economic Feasibility

Increasing the use of wind energy in Germany has been stimulated to a major extent by the introduction of a fixed feed-in tariff and the accompanying planning security through the Electricity Feed Act (1991 to 2000) and through the Renewable Energy Sources Act (since April 2000). Thanks to further technical developments and to economies of scale, the costs of wind turbines have been considerably reduced. At present, a machine with 2 MW of rated power, 90 m rotor diameter and a hub height of 105 m costs about

2.2 million ex works, with additional infrastructure costs of 25-30 % at the wind farm. At a reference site near the coast (with an annual average wind velocity of 5.5 m/s at an altitude of 30 m), about 6.1 GWh per year can be generated. This is sufficient to supply 1750 four-person households.

More important than the pure investment costs are the specific generating costs per kilowatt hour produced. Figure 7 shows an inflation-corrected reduction of the turbine costs per kWh produced annually at a reference site by 53 % between 1990 and 2004. This development corresponds to a learning curve with a rate of progress of 90 %. That is, for each doubling of installed power, the costs fell by 10 %.

While in 1991, the feed-in tariff amounted to up to 18.31 ct/kWh, by the year 2006 it had been reduced by 59 % to an average value of 7.44 ct/kWh. This historic development is extrapolated in the current Renewable Energy Sources Act (EEG), and is to be regularly reappraised. The minimum compensation for plants to be commissioned in the coming year is nominally 2 % lower. Taking inflation into account, new wind turbines therefore have to be approx. 4 % more cost effective per annum. This is especially challenging, considering how few available sites with high wind potential and without planning constraints exist anymore, and also the increasing prices of raw materials such as copper and steel.

Another provision of the EEG takes the importance of the local wind conditions for economic operation into account. This determines the amount and the stepwise digression of the feed-in tariff during the 20 year duration of the power purchase agreement. Clearly, economically ineffective projects have in the meantime been excluded from this incentive system. On the other hand, especially favourable conditions apply to offshore sites and the so-called repowering, i.e. the replacement of older, smaller machines by newer, larger turbines.

FIG. 7 | TURBINE COSTS

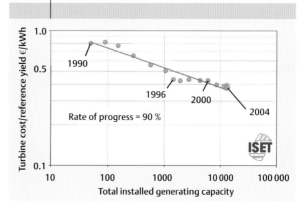

Time development of the turbine costs relative to the annual energy yield at a reference site, as a function of the total installed generating capacity (Graphics: ISET).

Nature Protection and Public Acceptance

Wind energy, with its industrial exploitation in the form of large wind farms, has lost the charm of historic windmills or of being an exotic alternative energy source. Wind power installations have however only local and minor negative environmental effects. They have to be compared with other interventions into Nature, such as the increasing concentration of CO_2 and other pollutants in the atmosphere, air and ground traffic, overhead transmission lines and many others. In view of the directly perceptible consequences of traditional energy supplies, a clear majority of German citizens are still in favour of the continuing development of wind power. Nevertheless, a paradoxical behaviour is often observed, accurately characterised by the sentence, "Not in my backyard!" (NIMBY); i.e. wind power *yes*, but somewhere else.

Considering specific wind farm projects, one can see the importance of planning which is socially and environmentally consistent, and which includes the interests of the local population and takes recognised minimum standards for nature and landscape protection into account. This can avoid political prejudices and polarisation on all sides, that are all too often observed and which can hardly be countered by citing scientific facts or technical solutions.

Ecological and Economical Expediency

With the climate change looming in the background, electricity producers are faced with a dilemma. In the coming decades, a major portion of the power generating capacity must be renewed. A social consensus concerning the new construction of nuclear power plants is not to be expected. The remaining options for power generation from fossil energy sources are hardly convincing, whether it be the exploitation of the still rich coal reserves in combination with technically immature and economically questionable CO_2 separation, or a politically risky, in the medium term expensive, and only palliative CO_2-reducing power generation based on imported natural gas.

For the renewable energy sources, in contrast, further cost reductions can be expected. New challenges are present, for example the integration of renewable sources into the power grid, as described above for the case of wind energy, and the transition of the power industry as well as the administrative and legal framework towards a more dispersed and intermittent power generation. The technical and economic perspectives are conspicuously improved when one takes the international energy supply system into account.

The ISET Institute of the University of Kassel demonstrated recently how the electrical energy supply of all of Europe and its neighbors could be secured using exclusively renewable energy sources and currently available technologies, at prices very close to those presently in effect. The central element of such a concept, with a very high proportion of wind energy, is the compensation of generation variations from the renewable energy sources among each other. This can be achieved by using a combination of different energy sources and by transporting energy through a transcontinental power network based on high-voltage direct current transmission (HVDC).

Along an ecologically and economically reasonable path to a largely renewable electrical energy supply, there are not only solvable technical hurdles to be overcome. A corresponding commercial-industrial and not lastly political transformation process on the part of the international energy suppliers and their regulators is equally necessary.

Summary

The rapid increase in the utilization of wind energy within the past fifteen years was made possible to a large extent by technological developments and a favourable political climate. Alongside the continued improvement of efficiency and economic competitiveness of the wind energy systems, political aspects are now becoming more important. Among these are integration into the national and international power grid and into the inter-national energy economy, as well as a societal consensus concerning energy policy. Power generation from wind energy is thus in transition from an alternative to a mainstream energy source. It can make a decisive contribution in the future to a climate-compatible and economically feasible power generation system.

References

[1] R. Gasch, J. Twele (ed.), Wind Power Plants – Fundamentals, Design, Construction and Operation, James & James, 2002.

[2] Institut für Solare Energieversorgung (ISET), Windenergie Report Deutschland 2005, reisi.iset.uni-kassel.de.

[3] German Energy Agency (dena), Integration into the national grid of onshore and offshore wind energy generated in Germany up to the year 2020 (dena Grid Study), 2005, www.dena.de.

[4] German Wind Energy Association (BWE), Press release 18/9/2006, www.wind-energie.de.

[5] B. Lange, Wind Power Prediction in Germany – Recent Advances and Future Challenges, European Wind Energy Conf., Athen 2006.

About the Author

Martin Kühn, born in 1963, studied physics engineering in Hannover, Berlin and Delft. Until 1999, he was a research assistant at the TU Delft, then through 2003 Project Manager for Offshore Engineering at GE Wind Energy GmbH. In 2001, he completed his dissertation at the TU Delft, and since 2004 he has held the first German Chair for Wind Energy at the University of Stuttgart, endowed by the Putzmeister company, Aichtal.

Contact:
*Prof. Dr. Dipl.-Ing. Martin Kühn,
Endowed Chair for Wind Energy (SWE),
Allmandring 5B, D-70550 Stuttgart, Germany
kuehn@ifb.uni-stuttgart.de*

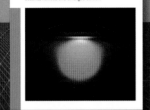

Hydroelectric Power Plants
Flowing Energy

BY ROLAND WENGENMAYR

Hydroelectric plants generate nearly one-sixth of the electric power produced worldwide. Water power is thus the only renewable energy source that contributes at present on a large scale to the supply of electrical energy for the world's population. It is efficient, but it can also destroy whole regions, societies and ecological systems.

Since the late 19th century, water power has been used to generate electricity. In the year 2003, according to the OECD, it provided around 16% of the world's total electrical energy requirements. It is roughly on a par with nuclear power [1,2]. Its contribution to the world's consumption of primary energy, which also includes heat energy, was 2.2% in the year 2003 [1]. It is thus the only renewable energy source which presently contributes an appreciable portion of the energy supply to the world's population.

Modern hydroelectric plants attain a very high efficiency. Up to 90% of the kinetic energy of the flowing water can be converted to electric power by modern turbines and generators. For comparison: light-water reactors convert only about 33 percent of the nuclear energy into electrical energy, while the remainder is lost as "waste heat"; coal-fired power plants have an efficiency of over 40 percent, while a modern natural gas combi-power plant achieves over 60 percent.

River and Storage Hydroelectric Plants

The hydroelectric power primer tells us that there are river power plants, storage power plants and pumped storage power plants. River power plants are used as a rule to supply the base load to the power grid. Their power production depends on the water level in the river, and this varies only slowly in most rivers over the seasons. Storage power plants are usually located high in mountainous regions and collect the water from melting snow in their reservoirs; they are therefore strongly dependent on this seasonal water supply. On the other hand, they can be started up within minutes and can be used to level out variations in grid power. They are thus well suited for compensating peak demand.

Pumped storage power plants, in contrast, are pure energy storage facilities which do not contribute to the production of electrical energy as such. In periods of low demand, they pump water into a high-level reservoir using power from the grid. As required at times of peak demand, they convert the stored potential energy back into electrical energy, by letting the stored water flow back down into the lower storage reservoir – or into a river. They are usually outfitted with special turbines which can work in the reverse direction as pumps. These plants are often used by the power companies for power 'upgrading': they are turned on when power is in short supply and therefore expensive.

The most modern pumped storage power plant in Germany went online in 2003 in Goldisthal, Thuringia. Its turbines were supplied by Voith Siemens Hydro Power Gen-

FIG. 1 | TURBINE TYPES

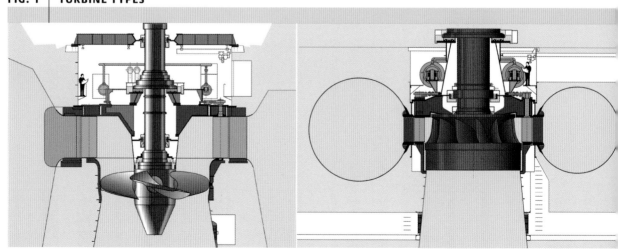

Renewable Energy. Edited by R. Wengenmayr, Th. Bührke. Copyright © 2008 WILEY-VCH Verlag GmbH & Co. KGaA, Weinheim. ISBN 978-3-527-40804-7

eration. This company, located in Heidenheim, is one of the world's major producers of equipment for hydroelectric power generation. Its founder, Friedrich Voith, constructed his first water turbines in 1870. In the year 2000, the turbine producers Voith and Siemens combined their hydroelectric power divisions in a joint venture, in order to hold their ground in the highly competitive world market. According to their own statistics, the combined enterprise has delivered turbines and generators for more than a third of the worldwide installed hydro-electric generating capacity.

Technical director Siegbert Etter explains that there is no other mechanical device which attains such a high power density as a modern water turbine. A turbine with a power output of 100 kilowatts is only 20 cm (8 in) in diameter, and is thus much more compact than an automobile engine of comparable power output.

How much energy a turbine can deliver depends essentially on the velocity and the amount of water that flows through it.

Large-scale Hydroelectric Plants

A typical flow power plant in a river without a significant head of water accepts a large volume of water at a relatively sluggish velocity. It uses Kaplan turbines, which are reminiscent of enormous ship's propellers (Figure 1, left side). At low rotational speeds, they extract the optimal amount of useful energy from the low water head at moderate flow velocities. The plant operators can adjust the pitch of the turbine blades and the fin-like guides in the housing through which the water flows into the turbine.

As the water head becomes higher, its kinetic energy also increases. Power plant owners therefore take advantage of the differences in altitude in mountainous regions, where storage reservoirs collect melt water at high levels. It flows down hundreds of meters through shafts and pipes to the power plant, where it jets out of nozzles into the massive buckets of Pelton turbines (Figure 1, right side). These modern descendants of the water wheel have dividing partitions in the centers of their buckets, which split the wa-

The Three-Gorges dam in China during construction in 2003. Behind its walls, which are up to 175 m (574 ft) high, the water of the Yangtze River will back up to form a lake more than 600 km (370 mi) long. (Photo: Voith Siemens Hydro Power Generation.)

ter jets as they hit the turbine. These jet splitters deflect the two halves of the buckets by nearly 180°, allowing the turbine wheel to extract nearly all the kinetic energy from the water. With a head of 1000 m, the water bursts out of the nozzles at nearly 500 km/h (more than 300 mph) and drives the turbines up to 1000 rpm. The nozzle openings can be adjusted by cones and a pivoted flow deflector can interrupt the flow of water to the turbines.

The largest power plants are built along the Earth's greatest rivers (Figure 2). Their massive dams do not produce a very high water head, but their turbines handle extremely large volumes of water. The controversial Three Gorges project currently under construction in China will be the largest hydroelectric plant in the world. Its dam is up to 175 m (nearly 600 ft) high and will back up the Yangtze River to form a lake almost 600 km (375 mi) long by 2008. The Three Gorges power plant is expected to become fully operational in 2009. The 26 giant turbines will then deliver 18.2 gigawatts of electric power. This corresponds to 14 nuclear power plant blocks or 22 large coal-fired plants [3]. The Francis turbines were designed by Voith Siemens

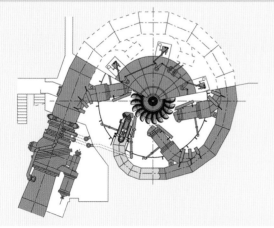

Left: The water (blue) flows horizontally into this vertical-axis Kaplan turbine past the control vanes (green). Center: Francis turbines are usually mounted with their axis vertical. The water flows in radially past the control vanes (green) and exits axially down the "outlet pipe". Right: A Pelton turbine (red) with an input pipe (penstock) leading to six steerable nozzles (one shown in cross-section); to the right of each nozzle is a flow deflector. A portion of the housing with mechanical controls is indicated schematically. (Graphics: Voith Siemens Hydro Power Generation.)

FIG. 2 | A LARGE-SCALE RIVER POWER PLANT

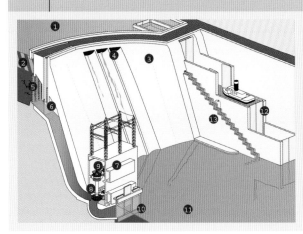

In a large-scale river power plant, the water at the upper level of the river (1) fills a reservoir. In order to allow controlled runoff of high water (higher water level (2)), the dam (3) has a spillway (4). In normal operation, the water flows past grids which can be raised to catch flotsam (5) and floodgates (6) through "penstocks" down to the powerhouse (7) and the turbines (8). The vertical-axis Francis turbines drive the generators (9) via shafts. The water then exits through lower floodgates (10) to the lower river level (11). The floodgates (6) and (10) are water-tight when closed, so that each turbine can be emptied of water and inspected. A system of locks (12) allows ships to pass the dam. The fish ladder (13) attracts the fish with a current of water and encourages them to choose this safe route.
(Graphics: Esjottes/von-Rotwein, Illustration + Infographic.)

Power Generation. Each of them is 10 meters (approx 33 ft) in diameter and weighs 420 tons.

Francis turbines can accept a large range of water velocities corresponding to moderate up to very high water heads, which are the domain of the Pelton turbines. Their curved buckets are not adjustable (Figure 1, center). The water flows through a delivery pipe (spiral) radially into the turbine and causes it to rotate. The water then exits downwards along the turbine axis through the outlet pipe to the lower water level of the river. Regulation is provided by the control surfaces arranged around the perimeter of the turbine wheel, whose jets are adjustable. The giant Francis turbines of the Three Gorge project operate at 75 rpm. With a water head of 80 meters (approx 260 ft), the massive water columns of the Yangtze flow into the turbines at a velocity of 20 km/h (approx 12 mph). The water is accelerated up to 120 km/h (approx 75 mph) on the rotating turbine buckets.

Modern Francis turbines extract almost all of the kinetic energy from the water and produce a large drop in pressure. This reduces the pressure at the outlet so drastically that the water foams up in cold water-vapor bubbles. This 'cavitation' has to be taken into account by the engineers when designing the turbines, since it must never be allowed to come into contact with the turbine blades. If the bubbles touch the metal, they implode violently and produce cavities in the surface of the blades (thus the term 'cavitation'). The turbines must be constructed and operated in such a way that cavitation occurs only at their outlets.

Each power plant has its own unique characteristics, and water turbines are tailor-made for a particular plant. The firm in Heidenheim currently designs them using complex computer simulations, and optimizes the design using small-scale models in their own test bed. This is itself a small power plant with a megawatt of output power. Ecological considerations can also influence the design of the turbines. In the USA, some power plants employ turbines which blow air into the water through special channels, thus increasing the low oxygen content of the river. There are even 'fish friendly' turbines: fish which have missed the fish ladder (Figure 2) have a chance of survival when passing through them.

Small-scale Hydroelectric Plants

Fish can also be an issue even for the constructors of small power plants. In order not to endanger the fish population of the small Black Forest River Elz, the power company WKV AG in Gutach constructed an elaborate inlet structure. A fish ladder and a fine grid keep the fish from entering the kilometer-long pipe which carries water parallel to the river into the turbines of the heavy machine factory. The factory obtains its electrical power to a large extent from the water of the Elz. The two Francis turbines with a total power output of 320 KW produce more power in the course of a year than WKV needs for its production lines. They deliver the excess power to neighboring households.

WVK supplies the market for small and medium water-power plants. The firm was founded in 1979 by a teacher, Manfred Volk, as a 'garage operation'. It has been growing ever since, and has delivered plants to over 30 countries. In mid-2005, the firm employed 65 engineers and technicians. Its turbine technology is developed by WKV in cooperation with the Technical University in Munich.

This Breisgau firm is successful, but it has to deal with the vagaries of the regenerative-energy market. According to WKV's financial director Thomas Bub, 70 to 80 percent of the projects planned by potential customers fizzle out for lack of financial backing. The particular economic aspect of water power lies in its extremely long useful life. Some plants are in use for more than 80 years. They can take full advantage of the cost-free energy supply over this long time period. On the other hand, the initial investment is often considerably higher and more elaborate than for a comparable fossil-fuel power plant. The interest payments on the high capital investment at the outset mean that many plants are amortized only after several decades. Thus, hydroelectric power needs investors and creditors who think in long terms.

Large Dams and their Consequences

This is particularly true of the billion-dollar investments required for large-scale hydroelectric plants. Their negative image has dampened the willingness to provide the investments on the part of traditional major backers such as the World Bank. Hydroelectric power on a large scale always exacts a high price. It can damage or even destroy whole regions, ecosystems and social structures. The World Commission on Dams (WCD) listed 45,000 large dams in the year 2000 [4]. Half of these dams were constructed for power production.

The WCD estimates that for the construction of these dams, worldwide 40 to 80 million persons were displaced or forced to move [4]. A famous negative example is the 50-year-old Kariba dam in Zimbabwe. It caused massive changes in the delta of the Zambezi River. 60,000 persons were forced to move by the construction of the reservoir [4]. In the case of the Chinese Three Gorge project, more than a million people must have been relocated, and the government of China has apparently not shown much consideration for their needs.

This policy has generated massive criticism on the part of non-governmental organizations such as the International Rivers Network. They have in the past applied political pressure to the World Bank so successfully that it practically withdrew from financing large hydroelectric projects in poorer countries. However, new financial sources have appeared which have again stirred up activity in this area. Especially India and China have offered partnerships to countries which are poor in capital but rich in water. They have shown fewer ethical scruples in their financing agreements and offer to deliver the technology at favorable prices at the same time. This will force the World Bank to make another policy reversal, if it wants to avoid being left out [5].

The safety of dams is another problem which must be taken seriously. In 1975, the Banquiao dam in China burst after a typhoon which was accompanied by catastrophic rainfall, and caused 26,000 fatalities. This was the greatest man-made catastrophe in history, says Stefan Hirschberg from the Paul Scherrer Institut in Villigen, Switzerland. But he also points out that in the OECD countries there has been only a single accident since 1969, which occurred in the USA and caused only a few deaths.

Hirschberg carries out systems research on energy technologies and knows the situation in China well. From his point of view, there are many advantages to not only small-scale but also large-scale hydroelectric power. A major plus point is the (in many cases) low emission rate of greenhouse gases. Reservoirs can – depending on their geological and climatic situation – emit carbon dioxide and, as a result of the decomposition of plant material, also methane. But even when the production of the materials for constructing a typical hydroelectric plant is taken into account, on the average only the equivalent of a few grams of CO_2 per kilowatt hour are released, Hirschberg explains. An average coal-fired power plant, in contrast, emits a kilogram (approx. 2 lb) of CO_2 per kilowatt of power produced, and a Chinese power plant emits even as much as 1.5 kg (approx. 3 lb) per KW.

The outdated Chinese power plants also lack filtering systems. In densely-populated areas, they shorten the lifespan of the population measurably. According to Hirschberg's studies, 25,000 years of life expectancy are lost there per gigawatt of power produced per year. Furthermore, the coal mines degrade the overall ecological and social conditions in China. They emit enormous amounts of methane, and only between 1994 and 1999, more than 11,000 miners lost their lives in accidents [6].

These statistics for a country with 25% of the world's population make it clear that large-scale hydroelectric power plants may be the lesser evil. Hirschberg in any case states, "In the context of global climate policy, hydroelectric power occupies an excellent position".

Summary

Water power generates nearly one-sixth of the electric power produced worldwide and thus defies the competition of all the other forms of renewable energy. With a low head of water, Kaplan turbines are suitable for electric power generation, while Francis turbines are used for moderate hydraulic gradients and Pelton turbines for very large gradients with high flow velocities. There are river power plants and storage power plants. Pumped storage power plants are used purely for energy storage. Modern water turbines and generators can convert the kinetic energy of the water into electrical energy with up to 90% efficiency. However, large-scale power plants can destroy whole regions, societies and ecosystems.

References

[1] International Energy Agency, Key World energy Statistics **2005**, 6.
[2] OECD Factbook 2005, Economic, Environmental and Social Statistics, available at lysander.sourceoecd.org
[3] www.voithsiemens.com/vs_en_references_south_east_asia_thrgorg.htm
[4] See www.irn.org/wcd/#sum (World Commission on Dams, Report Nov. 2000)
[5] Henry Fountain, "Unloved, but not Unbuilt", The New York Times **2005**, June 5.
[6] S. Hirschberg *et al.*, PSI Report No. 03-04, Paul Scherrer Institute, Villigen 2003.

About the Author

Roland Wengenmayr is editor of the German physics journal "Physik in unserer Zeit" and a science journalist.

Contact:
Roland Wengenmayr,
Physik in unserer Zeit,
Konrad-Glatt-Str. 17,
D-65929 Frankfurt am Main, Germany.
Roland@roland-wengenmayr.de

Concentrating Solar Power Plants

How the Sun gets into the Power Plant

BY ROBERT PITZ-PAAL

Concentrating solar power plants generate nearly as much power as all the photovoltaic modules coupled to power grids throughout the world. In principle, they work like big magnifying glasses: they collect the sunlight and concentrate it onto a thermal power generating unit.

Today, when people talk about solar power, they usually mean power produced by the shiny blue photovoltaic cells on the roofs of houses or along motorways. It is practically unknown that solar thermal power plants, which are based on a completely different operating principle, feed about the same amount of electric power worldwide into power grids, i.e. about 500 GWh per year [1]. The origins of this technology are in Europe. It is now advancing rapidly there, since Spain and Italy have begun subsidizing its commercial exploitation. Germany, Austria and Switzerland are too far north to be able to operate solar thermal power plants economically. Nevertheless, German research organizations and firms in particular are contributing intensively to further development of solar thermal power generation technology for the export market. In the future, the import of solar-thermally generated electric power could also become an important factor for the northern industrial countries in helping to reduce their CO_2 emissions [2,3].

The Principle

Over 80 percent of the electric power that we use comes from fossil fuel or nuclear power plants. The principle of power generation is in all ca-

ses the same: Heat energy from combustion of fossil fuels or from nuclear fission is used to drive a thermal engine – in most cases using steam turbines – and to produce electric current via generators coupled to the turbines. Solar thermal power plants use exactly the same technology, which has been refined for more than a hundred years. They simply replace the conventional heat sources by solar energy. In contrast to fossil energy sources, solar energy is not available around the clock. The gaps, for example at night, can be bridged over in two ways by the power plant operators: either they switch to fossil fuel combustion when the Sun is not available, or else they store the collected heat energy and withdraw this stored heat as needed for power generation. Solar thermal power plants work in principle like a magnifying glass. They concentrate the rays of the Sun, in order to obtain a high temperature: at least 300 °C (570 F) is required in order to be able to generate power effectively and economically with their thermal engines using the collected solar energy. The flat or vacuum-tube collectors familiar from rooftop applications are not suitable. The required high working temperature necessitates strong, direct solar radiation, and this determines which locations are appropriate for solar thermal power plants. They can therefore be operated economically only within the – enormous – Sun Belt between the 35th northern and 35th southern latitudes. This distinguishes them from photovoltaic systems, which can generate power effectively even with diffuse daylight, and are therefore also suitable for use under the conditions in Central Europe.

The Concentration of Light

If a black spot is irradiated by sunlight, it will heat up until the thermal losses to its environment just compensate input of solar radiation energy. When useful heat is extracted from the spot, its temperature will decrease. To reach higher temperatures, there are two possibilities, which can be used in parallel: reduction of the thermal losses, and an increase in the radiation energy input per unit area. The latter requires concentration of the direct solar radiation, which can be accomplished by using lenses or mirrors. But how strongly can solar radiation be concentrated?

The solar disc has a finite size; from the Earth, its diameter appears to us to correspond to an angle of about one-

Renewable Energy. Edited by R. Wengenmayr, Th. Bührke. Copyright © 2008 WILEY-VCH Verlag GmbH & Co. KGaA, Weinheim. ISBN 978-3-527-40804-7

Parabolic trough collectors concentrate the solar radiation and produce heat at a high temperature (400 °C (750 F)). The photo shows a section of the world's largest solar thermal power plant in California.

half of a degree. For this reason, not all the Sun's rays which reach the Earth are precisely parallel to one another. However, that would be necessary in order to be able to concentrate them onto a single point. Therefore, the maximum possible concentration is limited to a factor of about 46,200-fold. Nevertheless, we can in this way theoretically arrive at about the same radiation energy density as on the surface of the Sun, and could in principle obtain heat at a temperature of several thousand Kelvins. The focus point of the concentrator has to be at the same place during the whole day; the concentrator thus has to follow the Sun by moving along two axes. An alternative is offered by linear concentrators, for example cylindrical lenses: They do not concentrate the radiation at a single point, but rather along a caustic line, so that they need to be moved around only one axis in order to follow the Sun. In this case, however, the theoretical maximum degree of concentration is only 215-fold. This is still sufficient to obtain useful heat at a temperature of several hundred Kelvins.

Concentrating Collectors

In practice, mirror concentrators have for the most part taken predominance over lens concentrators. They are more suitable for assembly on a large scale and are less costly to construct. Essentially three different structural shapes can be distinguished (Figure 1). The **dish concentrator** is an ideal concentrator which follows the motion of the Sun along two axes. It consists of a parabolic silvered dish, which focuses the radiation of the Sun onto a single point. The receiver for the radiation, and often a thermal engine which is directly connected to it, are mounted at the focus point, both fixed in relation to the dish, so that they move with it. Wind forces, which deform the surface of the concentrator, limit its maximum size to a few 100 m² (a few 1,000 sq ft) and its electric power output to a few 10 kW. The **central receiver system** solves this problem by di-

viding up the oversized parabolic concentrator into a set of smaller, individually movable concentrator mirrors. These **heliostats** are directed onto a common focus point at the top of a central tower ('tower power plant'). There, a central receiver collects the heat. Since such a concentrator is no longer an ideal paraboloid, the maximum possible concentration factor decreases to 500- to 1,000-fold. This, however, is sufficient to reach temperatures up to 1,500 K (2,250 F). Large central receiver systems with thousands of heliostats, each with 100 m² mirror area, would require towers up to 100-200 m (330-660 ft) high. They could collect several hundred MW of solar radiation power. The **parabolic trough concentrator** is a linear concentrator which is moved along only one axis. A parabolic silvered trough concentrates the solar radiation up to 100-fold onto a tube which runs along the caustic line, and in which a heat-transfer medium is circulated. A similar design called linear fresnel collector uses a series of long, narrow, shallow-curvature (or even flat) mirrors to focus light onto a linear absorbers positioned above the mirrors.

FIG. 1 | CONCENTRATION OF THE SUNLIGHT

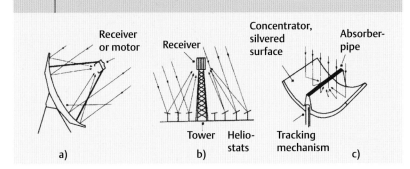

Three different possible methods of concentrating the solar radiation: a) a dish concentrator, b) a central-receiver system, c) a parabolic trough.

FIG. 2 | SOLAR THERMAL POWER PLANT

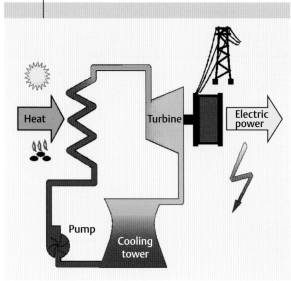

Heat

Turbine

Electric power

Pump

Cooling tower

High-temperature heat energy from the Sun or fuel combustion drives a steam turbine.

The theoretical maximum value of 215-fold concentration is not attainable in practice for two reasons: on the one hand, the large troughs "lie" on the Earth's surface and therefore cannot be rotated around all possible spatial axes to point towards the Sun, and on the other, surface imperfections reduce the geometric quality of the mirrors. Trough collectors can be joined up into lines many hundred of meters in length. Numerous parallel lines can then collect hundreds of MW of thermal power.

Power Cycles

A thermal power process can unfortunately not convert all the heat energy provided to it into mechanical work. It follows from the Second Law of Thermodynamics that part of the heat energy must be extracted from the process at a lower temperature than the input heat (so called "waste heat"). The higher the input temperature and the lower the output temperature, the larger the fraction of heat which can be converted into mechanical work (and thus into electrical power in a power plant). It therefore follows from thermodynamics that a high "temperature head"

between the hot and the cold heat reservoirs is more favorable for the thermal efficiency than a lower one.

In conventional thermal power processes to which solar energy can be applied, steam cycles (Clausius-Rankine process) are very often used:

Water is vaporized at high pressure in a boiler and the steam is further superheated. This hot steam expands in a turbine and performs mechanical work there. It is then condensed back to liquid water in a condenser and flows back into the boiler (Figure 2). The cooling of the condenser removes a part of the thermal energy from the process cycle and thereby fulfils the laws of thermodynamics.

Modern steam power plants operate at steam pressures of well above 100 bar (76 ktorr) and at temperatures of up to 600° C (1,100 F). As a rule, they generate an electric power of several 100 MWe. As a result, especially parabolic-trough concentrators and central-receiver systems are suitable sources of heat for this type of power plant, while dish concentrators can be used to drive other, more compact thermal engines with lower power outputs.

The solar energy concentrators in use today cannot quite reach the extreme steam temperatures and pressures mentioned above. The levels they reach nevertheless permit reasonable and efficient steam generation, if the steam power plant is designed to suit them. Since central-receiver or dish systems can in principle produce notably higher temperatures, it makes sense to utilize this potential. Due to the higher temperature of the input heat, the power plant can convert more heat into electrical power per unit area of its mirrors. As a result, for the same power output it requires less concentrator area, which saves on costs for its construction and operation.

Parabolic-Trough Power Plants

Parabolic-trough power plants were the first type of solar thermal power plants to generate electric power on a commercial basis. As early as 1983, the Israeli firm LUZ International Limited closed a contract with the Californian energy supplier Southern California Edison (SCE) to deliver power from two parabolic-trough power plants called SEGS (Solar Electricity Generating System) I and II. By 1990, all together nine power plants were built at three different locations in the Mojave Desert in California, with an overall power output of 354 MWe and more than two million square meters of collector area. In order to be able to deliver power reliably during peak use periods, these power plants are allowed to supply 25 % of their thermal input power from combustion of natural gas.

However, since fuel prices did not rise as originally expected, but instead fell, it was not possible to build additional cost effective-power plants. The existing solar power plants continue in service and feed nearly as much electrical power into the grid as all the photovoltaic systems worldwide, as mentioned above. Only at the end of the 1990's did the situation again change to the advantage of solar thermal power. In particular, the Kyoto Protocol and the planned dealing in CO_2-emission rights again made solar energy interesting. Thus, new projects are currently being put in operation in the USA and Europe and several 100 MW are under construction worldwide In some developing countries projects are supported through the World Bank.

In Europe, for example, projects have begun for two power plants, each with 50 MWe output power, called ANDASOL I and II. The cornerstone for ANDASOL I was laid in July, 2006, and construction of ANDASOL II began a year later. In the nine SEGS power plants, all together three generations of parabolic-trough collectors (LS1 – LS3) are in service. They all use a thermo-oil as heat-transfer medium; it is heated on passing through the solar collector and then flows into a heat exchanger where steam is generated. The first generation is still limited to operating temperatures of around 300° C (570 F) in the solar circuit, and can therefore use low-cost mineral oil as a heat-transfer medium. In the following generation, synthetic oils were introduced, which allow an increase of the operating temperatures to nearly 400° C (750 F). This makes it possible to generate steam at a higher pressure and temperature and thus attain higher energy conversion efficiencies.

The collectors of type LS1 have an opening width of 2.55 m (8.37 ft) and a collector length of 49 m (160.8 ft); in the third generation, the corresponding dimensions are 5.71 m (18.73 ft) and 99 m (324.8 ft). The absorber pipe within which the heat-transfer medium circulates is made of steel and covered with an optically-selective surface coating. It absorbs the radiation from the solar spectrum effectively, but re-radiates only a small amount of radiation energy and thus keeps the heat losses to the environment to a minimum. To further reduce these losses, the steel pipe is encased in an evacuated glass tube. The facets of the mirrors are made of thick glass with a reduced content of iron, which would absorb the light and is therefore unwanted. The glass is silvered on its rear surface. In the meantime, German producers have further developed the collector design of LUZ under

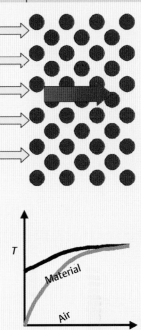

FIG. 5 | ABSORBERS

Above: Radiation (yellow arrows) is incident from the left onto a porous structure (red dots), while air passes through the structure from the left to the right; it takes up heat (blue-red arrow). Below: The dependence of the temperatures of the material in the structure and of the air as a function of the depth z within the structure.

<<< Fig. 3 *Prototype of the improved Euro-Trough Parabolic collector. (Photo: DLR.)*

<< Fig. 4 *The solar tower power plant CESA 1 at the European test Center 'Plataforma Solar de Almería' is currently a test platform for various new developments.*

< Fig. 6 *The European Dish-Stirling System namens EuroDish.*

FIG. 7 | DIRECT STEAM GENERATION

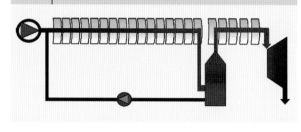

Direct solar steam generation in parabolic troughs: The water is partially vaporized in the first two thirds of the collector line. Then the mixture of steam and water is separated and the dry steam is further superheated in the last third of the line, while the hot water flows back to the inlet of the collectors.

the name EuroTrough (Figure 3). This new collector is lighter and stiffer, and is less costly to produce, to mount and to maintain. It was possible to increase the collector length to 150 m (490 ft) per tracking unit. An improved parabolic trough receiver pipe with an increased life time expectations and reduced heat losses developed by in Germany is also available. A complete collector circuit of the new design was tested in one of the SEGS power plants in California in parallel to the LUZ collectors before the technology is now implemented in the large scale in new solar power plants in Spain.

Central-Receiver Systems

Central-Receiver Systems are in the meantime also in commercial operation. A first plant with 11 MW output power (PS10) went online in early 2007. This can be considered a great success; since the beginning of the 1980's, around the world more than ten smaller demonstration plants with central receivers have been in operation (see Table 1 and Figure 4). Their operation was however terminated after the end of the test campaigns, since they were too small to be operated cost effectively.

The electric power was generated by a steam turbine in all these test plants. The main difference among the various test plants was in the choice of the transfer medium used to transport heat energy from the top of the receiver tower to the steam generator. It first appeared attractive to use the superheated steam itself as thermal transfer medium; this would eliminate the need for intermediate heat ex-

changers or steam generators and allow a direct connection to the steam turbines.

However, this concept soon showed two essential faults. In the first place, it was not easy to control the generation of superheated steam under conditions of fluctuating solar radiation input. The pressure and the temperature of the steam must be kept nearly constant in the turbine circuit. Secondly, with practicable technology it was nearly impossible to store heat energy within steam without considerable thermodynamic losses. Therefore, the recently erected commercial plant PS10 is based on saturated steam at moderate temperature and pressure to avoid these problems. In America, the concept of using molten alkali-metal salts as heat transfer media, which originated in France, was further developed and demonstrated between 1996 and 1999 at the 10-MW plant 'Solar Two' in Barstow, California. Mixtures of potassium and sodium nitrate salts can be optimized in terms of their melting points to the parameters required for steam generation. They offer two advantages: The relatively low-cost salts have good heat transfer properties; and furthermore, they can be stored at low pressure in tanks for use as a thermal storage medium. This makes it unnecessary to exchange heat with an additional storage medium. Their dis-advantage is their relatively high melting points, which, depending on the composition of the mixture, lie between 120° C and 240° C (250-480 F). This necessitates electrical heating of all the piping to avoid freezing out of the salts and resulting pipe blockage, for example during system start-up.

On the basis of experience gained with Solar Two, a Spanish-American consortium is planning to construct a larger successor plant in Spain under the name 'Solar Tres'. Solar Tres is expected to attain an output power of 15 MWe using a mirror area (solar field) increased by a factor of three, and its storage reservoir will be able to store sufficient energy for 16 hours of electric power generation.

The third concept makes use of air as heat-transfer medium. Air has, to be sure, rather poor heat transfer properties, but it promises simple manageability, no upper or lower limits to the operating temperature, unlimited availability and complete lack of toxicity. Air for the first time conjures up the vision of being able to operate combined gas and steam turbines at a high temperature using solar energy; these would make more efficient use of the collected solar heat energy, and thus of the mirror surface area.

In the first test setups, it was attempted to heat air by irradiating bundles of pipes through which it was passed. But only with the development of the so-called volumetric receiver has it become possible to sufficiently compensate the poor heat-

TAB. 1 | OVERVIEW OF CENTRAL-RECEIVER SYSTEMS WORLDWIDE (TEST PLANTS)

Project Name	Country	Power/ MWe	Heat transfer medium	Heat storage medium	Start of operation
SSPS	Spain	0,5	Liquid sodium	Sodium	1981
EURELIOS	Italy	1	H_2O steam	Nitrate salts/H_2O	1981
SUNSHINE	Japan	1	H_2O steam	Nitrate salts/H_2O	1981
Solar One U.S.A.	USA	10	H_2O steam	Oil/Stone	1982
CESA-1	Spain	1	H_2O steam	Nitrate salts	1983
MSEE/Cat B	USA	1	Nitrate salts	Nitrate salts	1983
THEMIS	France	2,5	Hitech Salts	Hitech salts	1984
SPP-5	Ukraine	5	H_2O steam	H_2O steam	1986
TSA	Spain	1	Air	Ceramic reservoir	1993
Solar Two	USA	10	Nitrate salts	Nitrate salts	1996
PS10	Spain	11	Saturated steam	steam	2007

transfer properties of the air. Such a receiver contains a 'porous' material, for example a meshwork of wire, which is penetrated by the concentrated solar radiation and through which at the same time the air to be heated flows (Figure 5). The large internal surface area guarantees efficient heat transfer. If the air circuit is open and operates at atmospheric pressure, then such a receiver can drive steam turbines. If, on the other hand, the air receiver is closed with a transparent radiation window and the air is pressurized, then the system can even be used with gas turbines.

Air systems at atmospheric pressure are practically free of operating disturbances, and this is the reason why they are favored by a European consortium: In 1994, at the Plataforma Solar a 3 MW test system operated. In the meantime, further research conducted by the DLR within the European Network was able to increase the efficiency of individual components and to reduce the costs of the receiver and the storage reservoir. A first demonstration project with 1.5 MW of electric output power is currently being set up in Germany and will serve as a technical benchmark for larger projects in the Sun Belt.

Dish-Stirling Systems

To date, Dish-Stirling systems have a prototype status with electricity generation costs still significantly above those of large-scale central receiver or parabolic trough power plants. Companies in the USA and in Germany are currently working on four different systems worldwide (Figure 6). The system which is furthest along in its development originated in Germany and has accumulated several tens of thousands of hours of operation.

Such systems aim at independent power generation, not coupled to a power grid, in the form of Dish Parks made up of several thousand individual systems. Projects of this type are currently in the planning stages in the USA. Their principal advantage is a very high efficiency of up to 30 %: this is provided by the combination of a nearly ideal paraboloid concentrator with an excellent thermal engine. If the Sun is not shining, then Dish-Stirling systems can in principle be operated with fuel combustion, in order to meet the demand for power. This is a decisive advantage over photovoltaic cells, which aim at a similar market: They, however, require expensive storage batteries for the same purpose.

These are good reasons why dish-Stirling systems have a favorable market chance in the medium term for independent power generating applications. For this purpose, they must be capable of autonomous and very reliable operation. Subsidized niche markets are, however, only one of the possibilities for dish-Stirling systems. A still greater market potential lies in the increasing power requirements of developing countries, especially those with a large amount of sunlight, poorly established power grids and high costs for the import and transport of fossil fuels.

Aside from technical maturity, the small number of units produced represents a hurdle to commercial marketing of dish-Stirling systems. However, cost may decrease in the near future, since Stirling Energy Systems and a Californian utility have signed a contract to provide grid power by using a park installation of dish/Stirling systems. A capacity of 500 MW (20,000 units) is intended for implementation starting in 2009 and to be completed in 2012.

Cost Effectiveness

In the research and demonstration systems of the 1980's, the costs of power generation were still in the range of 50 to 100 €-cents/kWh (approx. 70-145 $-cents/kWh). The SEGS power plants were the first to reduce these costs significantly with their commercial technology. In the first SEGS plants, they were around 30 cents/kWh; with technical improvements and upgrading, they sank to about 12 to 15 €-cents /kWh (17-22 $-cents/kWh).

The profitability of a solar thermal power plant naturally strongly depends on its location. The available solar energy influences the costs per kWh approximately linearly. At the SEGS sites in the Mojave Desert in California, annually about 2.5 times as much direct solar radiation is available as in Germany, and still 25 % more than in Southern Spain.

If one assumes the same conditions of irradiation and compares them with good sites for wind power plants, the result is that electric power from solar thermal power plants is at present about twice as expensive as power from wind plants and around half as expensive as power from photovoltaic cells. In computing the costs, one must distinguish between large installations with several 10s of MWe of electrical output power, and small applications which are not connected to the power grid. The numbers mentioned above hold for large plants and still contain considerable potential for cost reduction.

Power generators which are coupled to a power grid have to compete today in a deregulated energy market with conventional power plants, which can generate power for about 4 €-cents/kWh (6 $-cents/kWh) and even less. This is considerably less than in decentral markets, where regenerative power generation can be the cheapest power source – which simplifies the market entry of suitable regenerative technologies. However, in order to attain an order of magnitude with renewable energy generation which is commercially relevant, the power-grid market plays a decisive role. Therefore, the introduction of these technologies into today's market is being promoted by various organizations and agencies.

Technical Improvements

A clear-cut cost reduction can be expected from the following factors: automated mass production of a large quantity of components, an increase in reliability of the plants, and the extensive automation of their operation and the cleaning of the collectors. An important contribution is also promised by further improvements in the technology and innovative concepts for large solar-thermal plants. In Germany, these goals are being pursued at the German Aero-

FIG. 8 | AN IMPROVED TOWER POWER PLANT

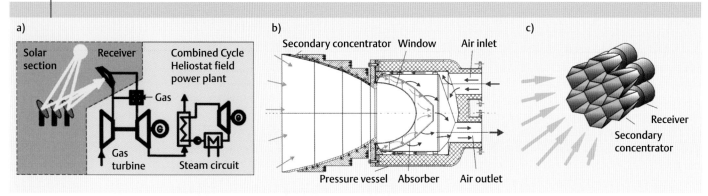

a) Schematic of a solar combi power plant; b) Layout of a high-temperature receiver module; c) The conical mirrors of a number of modules can together make effective use of the concentrated radiation even with a large focal-spot area.

space Center (DLR) as part of its energy research program, together with industrial partners. We will mention briefly some of these research activities here.

An important aspect is increasing the operating temperatures, which, as explained above, will improve the conversion efficiencies and permit a smaller specific collector area to be used. For parabolic-trough collectors, the operating temperature limit of the thermo-oils in use must be increased above 400 °C (750 F). One possibility, which has already been tested, is the direct evaporation and superheating of water in the collector itself (Figure 7). For this test, a 500 m (1,650 ft) long collector loop was set up at the Plataforma Solar in Almería. Among other things, the regulation behavior and flow properties of the water-steam mixture in absorber tubes is investigated with this apparatus. More than 7,000 hours of test operation have proven the technical feasibility of this concept. It should yield a decrease in the power-generating costs of around 10 %.

Also linear Fresnel Systems aim at lower overall costs because the aperture size for each heat absorbing element is not constrained by wind loads as in the case of parabolic troughs. Low cost flat glass can be used and curved elastically, due to the large curvature radius of the facets. The absorber is stationary so that flexible fluid joints are not required. Suitable aiming strategies permit a denser packing of mirrors on limited available land area. However, due to the flat arrangement of the mirrors intrinsic additional optical (cosine) losses reduce the annual output by 20 - 30 % compared to the parabolic trough design. This reduced optical performance needs to be offset by the lower investments. Prototypes of these systems have recently been built in Australia, Belgium and at the Plataforma Solar de Almería in Spain, but performance and cost data have not been published yet.

Intensive research is being carried out on central-receiver systems with the goal of using pressurized air as the heat-transport medium for solar energy, allowing a high input temperature for driving a gas turbine (Figure 8a). A decisive factor is using the right technology to transfer the

concentrated solar radiation through a glass window into the pressure vessel of the receiver (Figure 8b). Since the diameter of such heat-resistant quartz glass windows is limited by their fabrication process, a number of such modules are arranged in a matrix with conical mirrors (secondary concentrators) in front of their entrance windows. These mirrors are shaped so that together, they form a large entrance aperture with practically no gaps (Figure 8c).

In an experiment at the Plataforma Solar, up to now three such modules have been combined and connected to a small 250 kW gas turbine. They produce temperature up to 850° C (1,560 F) at a pressure of 15 bar (0.02 torr). In early 2003, the turbine generated electric power for the first time. This represented an important milestone on the way towards a large-scale technical application. The researchers expect a reduction of generating costs from this concept of up to 20 %.

An additional important component which can contribute to cost reduction is the thermal energy storage reservoir. When a solar thermal power plant is operated on solar energy alone, the operation period of the power generating block which it drives at a favorable site is equivalent to an annual full-time operation of up to 2,500 hours. This period considerably increased if it were possible to store the energy from the solar field in a cost-effective manner. Then, the power plant could be equipped with a second collector field of the same size as the first, whose collected solar energy would flow into the storage reservoir. In times with little or no sunlight, the power generating block would then make use of this stored energy.

This increase in operation period would save the cost of investment for a second power generating block. The precondition is of course that the costs of the thermal energy storage reservoir are less than the additional cost of a larger power generating block. From present knowledge, this is possible. Cost-effective thermal energy storage concepts promise a further reduction in power-generating costs, which could again be up to 20%. Such a thermal energy storage reservoir would also have additional advantages. With it, power could be generated according to grid requirements, i.e. at peak demand periods. The price paid per kWh is then highest. It is also a plus on the technical side that the power plant would always operate under optimal load conditions and could thereby minimize its heating-up and cooling-down losses.

The development of storage systems was long neglected in Europe: initially, the use of fossil fuels for bridging over periods with low sunlight was seen as the cheapest alternative at least as a first step. However, it has two disadvantages: firstly, many subsidy arrangements do not permit hybrid operation (for example laws governing the subsidized input of power to the grid). Secondly, the use of fossil fuels in a power plant which has been optimized for solar operation is ineffective and therefore economically unfavorable.

A system is currently under development for the parabolic-trough collectors with operating temperatures up to 400° C (750 F) which will permit intermediate storage of the heat energy in large blocks of high-temperature concrete. With central-receiver systems, depending on the heat-transfer medium, there are two types of storage reservoirs. One type uses tanks containing salt melts; the other transfers the heat from air to a packed bed of small solid particles, which allow the air to pass between the particles and offer a large surface area, e.g. ceramic balls or quartz sand.

The Lowest CO_2 Emissions

Solar thermal power plants are an important intermediate link between today's energy supply based on fossil fuels and a future solar energy economy, since they incorporate important characteristics of both systems. They have the potential of supplying the world's electrical energy requirements several times over from solar fields, and by means of simple storage methods, they can potentially deliver power as needed, in contrast to other renewable energy sources such as wind energy.

Solar thermal power plants are also favored by the fact that they can reduce CO_2 emissions in a particularly effective way. This becomes clear if one sums up the emissions which are due to the fabrication of the components, construction, operation, and shut-down of the plant via life-cycle analysis. Comparing various renewable energy sources in this way, one finds the following balance for the specific CO_2 emissions per MWh of electrical energy generated: For solar thermal power plants, only 12 kg (26.5 lb) CO_2/MWh are emitted, while hydroelectric power plants emit 14 kg (39 lb), wind-energy plants emit 17 kg (37.5 lb), and photovoltaic power plants emit 110 kg (243 lb) CO_2/MWh.

Photovoltaic power generation is so unfavorable in this comparison because the production of the semiconductor modules still is very energy intensive and therefore produces a large amount of emissions. By comparison: modern gas and steam turbine power plants emit 435 kg (960 lb) of CO_2/MWh, and coal-fired power plants as much as 900 kg (1,980 lb) CO_2 per MWh generated. These emissions are mainly due to the combustion of the fossil fuels. For these reasons, different energy scenarios, for example that of the WBGU (the Scientific Advisory Council of the German Federal Government on Global Environmental Changes), predict that solar thermal power plants will contribute a significant fraction of the electric power consumed worldwide by the year 2050, up to 25% of the total. In the northern countries, this will be accomplished by importing electric power. Estimates indicate that worldwide, by the year 2025 around 40 GW of electrical generating capacity from solar-thermal energy will be installed.

Summary

Concentrating Solar power plants collect sunlight, like giant magnifying glasses, and use it to drive thermal engines for electric power generation. Three types of construction have established themselves at present. Systems with sun-tracking, silvered parabolic troughs which concentrate the solar radiation onto a central absorber tube, through which a heat-transfer medium flows, are already in commercial operation. In the central-receiver systems, a field of sun-tracking mirrors focuses the sunlight onto the top of a tower; a receiver there passes the heat energy to a thermal-transport medium. For small, decentral applications, dish-Stirling systems are suitable. These are sun-tracking, paraboloid mirror dishes with a Stirling motor at their focus points.

References

[1] Renewables Information **2003** — 2003 Edition, Publisher IEA, 201 pages (Jouve, Paris 2003).

[2] Trieb, Franz; Müller-Steinhagen, Hans 2007: Europe - Middle East - North Africa Cooperation for Sustainable Electricity and Water. Sustainability Science, **2007**, 2 (2)

[3] R. Pitz-Paal, Editor, Journal of Solar Energy Engineering **2002**, 124 (5), 97.

About the Author

Robert Pitz-Paal has carried out research on solar energy since 1993 at the German Center for Air and Space Travel (DLR) in Cologne-Porz. He is Division Leader for Solar Research there and Professor at the RWTH (Technical University) in Aachen, Germany.

Contact:
*Prof. Dr.-Ing. Robert Pitz-Paal,
Deutsches Zentrum für Luft- und Raumfahrt e.V.,
Institut für Technische Thermodynamik,
Solarforschung,
Linder Höhe, D-51147 Köln, Germany.
Robert.Pitz-Paal@dlr.de*

Photovoltaic Energy Conversion

Solar Cells – an Overview

BY ROLAND WENGENMAYR

The market share of photovoltaic power is growing rapidly. However, its contribution to the overall generation of electric power is still small. Photovoltaic systems are at present a mature technology and are long-lived, but they are still too expensive. New materials and production methods can be expected to change that.

Photovoltaic energy conversion is impressively elegant, since it transforms the energy of sunlight directly into electrical energy. Since the American inventor Charles Fritts constructed the first selenium cells in 1883 (see Table 1), solar cells have developed into a technology which is mature and reliable from the user's point of view. At present, silicon technology dominates the production of solar modules, with over 90 % market share. Alternative materials are only slowly gaining ground. Silicon is the standard material in the electronics industry, and its properties have therefore been thoroughly investigated, while its industrial process technology is well established.

Only rather recently have the solar-cell producers emerged from their niche in the shadow of the all-powerful chip manufacturers. With an increasing economic leverage, these young companies can bring new technologies to the market, which are more suited for photo-voltaic applications. The current fabrication methods have in particular very high losses of valuable semiconductor material. Wafers of monocrystalline silicon, the preferred material up to now, are the starting material for the production of solar modules. They are cut out of costly single-crystal blocks ("ingots"). With polycrystalline silicon, corresponding discs containing many small crystals are also cut out of silicon blocks. The blocks are either cast from molten silicon in crucibles, or else the material in the crucible is melted by induction heating using strong electromagnetic fields. The step of sawing out the wafers reduces much of the starting material to powder, both in the case of monocrystalline and of polycrystalline silicon. In the process, a considerable amount of energy is wasted, which was invested in growing the crystal or in melting the multicrystalline ingots. This worsens the energy balance and increases the production costs. Only new technologies and materials, which are presently in various stages of development from research through pilot projects to commercial processes, can solve this fundamental problem of the conventional photovoltaic industry.

In the production and application of photovoltaic modules, Germany, Japan and the USA are the three leading countries that represent 90 % of the total worldwide photovoltaics installations. In spite of growth rates of around 30 % per year in Germany, the biggest market in the world so far, considered in absolute terms, is still small. According to the figures of the German Federal Ministry for the Environment, Nature Conservation and Nuclear Safety, the total installed power in Germany in 2006 was 2831 megawatts peak power (MWp) [1]. These photovoltaic systems generated all together 2000 gigawatt hours of electrical energy in 2006 [1]. This corresponds to 0.4 % of the overall electric power consumption in Germany. In the USA, in 2005, 108 MWp in total were installed, 65 MWp grid connected [2]. Worldwide, the annual sales of photovoltaic technology have grown to more than 5 gigawatts peak power (GWp) [2].

Within the solar industry, by the way, the power of solar modules is usually defined in terms of watts/peak (Wp), or peak power. This corresponds to normalized test conditions and is used to compare different modules. It however does not necessarily represent day-to-day operating conditions, and also does not reflect the power produced by the module when the sunlight is at its strongest.

TAB. 1 | **HISTORICAL MILESTONES**

Year	Type of Solar Cell	Efficiency	Developer
1883	selenium (photocell)	nearly 1 %	Charles Fritts
1953/4	monocrystalline silicon	4.5-6 %	Bell Labs, USA
1957	monocrystalline silicon	8 %	Hoffmann Electronics, USA
1958	monocrystalline silicon	9 %	Hoffmann Electronics, USA
1959	monocrystalline silicon	10 %	Hoffmann Electronics, USA
1960	monocrystalline silicon	14 %	Hoffmann Electronics, USA
1976	amorphous silicon	1.1 %	RCA Laboratories, USA
1985	monocrystalline silicon	20 %	University of New South Wales, Australia
1994	gallium indium phosphide/ gallium arsenide, with concentrator	over 30 %	National Renewable Energy Lab (NREL), USA
1996	photoelectrochemical, Grätzel	11.2 %	ETH Lausanne, Switzerland
2003	CIS thin-film	19.2 %	NREL, USA
2004	multicrystalline silicon	20.3 %	Fraunhofer ISE, Freiburg, Germany
2005	aluminum gallium indium phosphide/ gallium indium arsenide/ germanium; three layers, with concentrator	39 %	Boeing Spectrolab, USA

Renewable Energy. Edited by R. Wengenmayr, Th. Bührke. Copyright © 2008 WILEY-VCH Verlag GmbH & Co. KGaA, Weinheim. ISBN 978-3-527-40804-7

Fig. 1 *The glass roof of the Stillwell Avenue station on the New York subway in Brooklyn contains one of the largest thin-film solar-cell roof top installations in the world, with a nominal power output of 210 kWp* (Photo: Schott).

Efficiencies are Still Low

Especially when compared to wind and hydroelectric power, photovoltaic power conversion is still very much in the background, in spite of considerable government subsidies. Owing to the waste of material in the currently-used production processes, it is up to now also inferior to the other renewable energy sources from the ecological point of view. This can be seen by taking a look at the "harvest factor": It represents the useful energy that a plant generates in the course of its life cycle in relation to the energy that was required for its construction and installation. This average harvest factor is in the range of 7 for a modern photovoltaic plant. With a life expectancy of around 20 years, the plant will thus have been amortized in terms of energy after three years of operation; thereafter, it will generate more technically usable electrical energy than was required for its construction.

Wind energy, in contrast, yields harvest factors between 10 and 50, while large hydroelectric plants, due to their long life cycles, have values up to as much as 250. In the case of coal-fired power plants, the harvest factor is about 90, and for nuclear power plants, it lies between 160 and 240; the energy cost of mining, transporting and consuming the fuel is included in these numbers. Many experts estimate, however, that photovoltaic systems will in practice have

lifetimes of up to forty years. Then their average harvest factor would increase to at least 13.

The energy costs, materials costs and – in absolute terms – a still small market share mean that the investment costs for photovoltaic systems remain high. These high costs, in turn, cause high "power-generating costs". In the most favorable case, they are today at the level of 50 -cents per kilowatt hour, as Giso Hahn discusses in more detail in his overview in the following chapter. Thus, a private installation on the roof of a house pays off only after many years, even with government subsidies.

In spite of this sober interim balance, photovoltaics offer enormous opportunities. Not only the threatening climate collapse provides a weighty argument in favor of strong continuing support for research and development of photovoltaic devices. Photovoltaics are still a young technology as a large-scale energy source, even though they were already used in space vehicles beginning in the 1960's, and they have a strong potential for further development. Solar cells can offer some tangible plus points in terms of decentralized energy supplies. Windows coated using the new thin-film processes can combine electric power production in an elegant manner with the necessary light and heat management, especially for large buildings. This development is still in its infancy. It is also probable that in

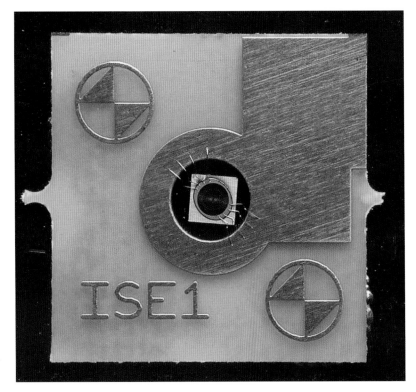

Fig. 2 *This little concentrator cell, with a gallium arsenide absorber, attains an efficiency of up to 35 %* (Photo: Fraunhofer ISE).

Fig. 3 *A FLATCON™ module from Concentrix Solar (Freiburg) with concentrator cells* (Photo: Fraunhofer ISE).

view of the increasing global energy demand, the price of electric power will continue to rise and this will make an improved version of photovoltaics economically competitive in the future.

Area-dependent and Area-Independent Costs

The costs of photovoltaic plants are composed of a contribution which is proportional to the area covered by the solar modules, and a contribution which is independent of the area. The latter includes for example the expensive power inverters, whose price has however been reduced in recent years by at least 20 % through new technical developments. For larger plants, which are to be connected to the power grid, the inverter converts the direct current from the solar modules, which are in a series circuit, into alternating current. Whoever wishes to use solar power to operate normal household appliances has to employ such a power inverter. If it is technically faulty or poorly matched to the solar modules, it can cause serious power losses and reduce the overall efficiency of the plant. In older installations, the power inverters were often a weak point. Modern units as a rule operate very reliably and, with proper installation, efficiently.

Among the area-dependent costs, besides the installation costs, the supporting frames and the cost of the land which may be a consideration, the price of the solar modules themselves is a significant factor. For this reason, researchers and developers are working steadily to further improve the efficiencies and reduce the fabrication costs of the solar modules. Progress is however difficult, although there is no lack of new concepts and materials combinations. The bottle-neck on the way to the market is the industrial processing. Production plants are expensive and take some years to be amortized, and they have to yield modules of reliable high quality. Therefore, many producers of solar modules are technically conservative. It requires considerable time before innovative processes reach commercial maturity. As a result, monocrystalline and polycrystalline silicon will continue to dominate the market for some time to come, although silicon, as an indirect semiconductor, has notable disadvantages (see the infobox "Brief Fundamentals of Photovoltaics" on page 37). Until 2002, wafers made from large silicon single crystals were still in the lead, i.e. the same expensive starting material as is used by chip producers. The solar industry makes use of their rejected batches, since photovoltaic devices are more tolerant towards crystal defects than highly integrated electronics devices. In 2002, wafers made of poly- or multicrystalline silicon took for the first time a larger market share than single-crystal material. This was the first material to be developed especially for photovoltaics applications.

Since both these starting materials for solar modules suffer from a high materials loss during processing, the industry has for some time been searching for new methods which would avoid sawing the wafers out of a block with an inevitable waste of material. These new processes are collectively referred to as silicon ribbon or silicon foil. The raw material for the modules is in this case pulled directly from the silicon melt in the required thickness. Giso Hahn describes these fascinating methods in the following chapter. Silicon ribbon has in the meantime captured its own small share of the market.

Modules fabricated from single-crystal silicon have thus far yielded the highest efficiencies. Commercial cells convert up to 20 % of the incident sunlight into electrical energy. Commercial multicrystalline modules yield at best 16 % efficiency; for silicon ribbon or foil, the efficiencies are – depending on the fabrication method – 13 to 18 %.

Thin Films for Glass Facades

A different approach is used in processes in which the active photovoltaic layer is deposited as a very thin film onto the substrate material, usually glass. Such films can be made for example of cadmium telluride or copper indium disulfide (CIS). On page 50, Nikolaus Meyer introduces a young company which is developing this CIS technology for the market. The German firm Würth Solar in Schwäbisch Hall started the first mass production of CIS solar modules worldwide in October 2006. Their production of 200 000 solar modules in January 2007 corresponds to an annual capacity of 15 megawatts of output power.

Another chalcopyrite material is a compound of copper, selenium and either indium, gallium or aluminum ($CuInSe_2$, $CuGaSe_2$ or $CuAlSe_2$). Scientists at the Hahn-Meitner-Institute in Berlin have already obtained efficiencies of nearly 20 % in thin-film solar cells based on chalcopyrites.

Silicon is also suitable for thin-film processing. The process consists of growing a thin layer of either microcrystalline or amorphous silicon onto a substrate material. Microcrystalline means, in comparison to multicrystalline, that a large number of microscopically small silicon crystallites make up the layer. In an amorphous solid, the atoms, in contrast, no longer exhibit a long-range spatial order over more than a few atomic radii. Amorphous silicon (a-Si) is enriched in hydrogen during its preparation process (a-Si:H). The hydrogen is able to partially "repair" the many defects in the disordered material and thus prevents the efficiency of the resulting solar cell from being reduced too seriously (see also the infobox "Brief Fundamentals of Photovoltaics").

Thin-film solar cells made from a-Si:H can be readily fabricated on relatively large areas of a glass substrate. Commercial modules currently have typical efficiencies around 8 %, if they are required to be partially transparent as a glass coating. Up to 12 % efficiency is possible if they have a reflective coating behind the absorber layer. Thin-film modules of a-Si have the disadvantage that their efficiency at first drops off by 10 to 30 % of its initial value, before it stabilizes after a certain operation period. On the other hand, they are much less sensitive to high operating temperatures than crystalline silicon solar cells. These lose about 0.4 % of their efficiency for every degree Celsius above their standard operating conditions (see the infobox "Determining the Efficiency" on p. 48); the efficiency loss of a-Si:H is in contrast only 0.1 %. As a result, if the efficiency of a crystalline silicon solar cell at 25 °C (77 F) is, for example, 18 %, then it is reduced at 50 °C (122 F) to 15,5 % – that is, under realistic operating conditions. A cell made of a-Si:H would lose only one-fourth as much. Modules made from crystalline silicon must therefore be effectively cooled. At very hot locations, a-Si:H can have a superior energy yield. Glass coated with a-Si has a lower efficiency, but for architectural applications, it has very attractive properties. The transparent modules can – along with power generation – also carry out the central function of active light and heat management.

BRIEF FUNDAMENTALS OF PHOTOVOLTAICS

When a photovoltaic cell absorbs a light quantum (photon) from sunlight, this photon can raise the energy of an electron out of the bound states ("valence band") in the solid. The electron leaves an empty "hole" in the crystal lattice. This hole can hop from atom to atom and thus can – like the excited electron – contribute to an electric current. To be sure, this works only when the electron is excited at least into the next higher allowed energy level (the "conduction band"). In between, in solids there is an energetically forbidden zone, the so-called energy band gap. The photon must therefore be able to give sufficient energy to the electron to cross this zone.

The flow of current within the solar cell is reminiscent of the two-storey Oakland Bay Bridge in San Francisco: on the upper storey the "electron traffic" flows in one direction, while on the lower storey, the ""hole traffic" flows in the opposite direction. The cell obtains electrical energy from this bidirectional flow. This can however function only when the two currents have predetermined flow directions to the corresponding electrodes, i.e. when the flows are strictly controlled along "one-way streets". This is achieved by two regions in the cell, which act as valves to allow the flow only in the desired direction: the "p-doped region" lets holes pass through in one direction to the electrodes, while the "n-doped region" passes electrons in the opposite direction. The solar cell thus works like a rectifier or diode. In fact, it reverses the action of a light-emitting diode, in that it converts the energy of light quanta into electrical energy.

Important for the fundamentals of solar cells is the difference between direct and indirect semiconductors. In direct semiconductors, such as gallium arsenide, the forbidden energy zone is sufficiently narrow that photons from sunlight can lift the electrons directly across it and into the conduction band. Silicon, in contrast, is an indirect semiconductor. In this case, coupled vibrations of the atoms in the crystal (phonons) have to help in order to raise the electrons into the allowed energy region. Silicon is therefore a less favourable material for photovoltaic applications. Only amorphous thin-film silicon is a direct semiconductor.

The excited electrons and holes must also have a long "lifetime", so that the electric current can flow in the cell with a minimum of perturbations. Defects in the crystal lattice allow some of the electrons and holes to "recombine" with each other; this is an unwanted side effect which allows them to be lost to the production of electrical energy. This is also the reason why amorphous silicon has not long since attained the same efficiency as crystalline silicon, although it is a direct semiconductor. In new, less perfect materials such as silicon ribbon, a sophisticated "defect engineering" is necessary; this is described by Giso Hahn in the following chapter.

A more detailed description of the physics of solar cells is given for example in [7].

Fig. 4 *This photoelectrochemical solar cell (dye solar cell) from the ETH Lausanne is – at 11 % efficiency and with a high thermal stability – on the way to giving serious competition to conventional silicon solar cells (pictured in the background).*
(Photo: CH-Forschung).

current world's largest thin film solar power plant (2007) "Rote Jähne" is located in Eastern Germany, near Leipzig, consists of more than 90,000 modules and has a total output capacity of 6 MWp. It provides enough electrical energy to power some 1,900 homes.

Highest Power Outputs

The attempt to increase the efficiencies of single-crystal silicon runs up against a theoretical limit of 28 % under standard conditions (see Giso Hahn's chapter). Another semiconductor material, gallium arsenide, permits a further increase, since it is a direct semiconductor (see the infobox "Brief Fundamentals of Photovoltaics"). Gallium arsenide is therefore often used in optoelectronics. It has, however, the disadvantage of being an expensive material. It also cannot be structured as readily as silicon using standard semiconductor techniques. Even with gallium arsenide, efficiencies above 30 % can be obtained only if the solar cells are fabricated with a complex structure.

An example is the tandem cell. A cell of this type, which has an especially high efficiency, was introduced by researchers from the Fraunhofer Institute for Solar Energy Systems (ISE) in Freiburg in 2005. Three layers of the materials gallium indium phosphide, gallium arsenide, and germanium are stacked in a sandwich structure. Each layer can convert a different portion of the solar light spectrum into electrical energy. The cell thus makes much more efficient use of the incident sunlight than a conventional solar cell. Furthermore, the Freiburg group put a lens above their tiny cells, which have an area of only 0.031 cm^2 (0.005 in^2), corresponding to the size of a light-emitting diode. The lens focuses the sunlight onto the small cell and concentrates it there by a factor of 500. The result is an efficiency of 35 %, which represents the European record. The world record is at present held by a concentrator cell from the Boeing spin-off Spectrolab in California, with 39 % efficiency – under special test conditions. The concentration of sunlight using lenses thus makes it possible to exploit even very complex and expensive cells optimally. This offers the possibility of being able to market such cells in the future at competitive specific-area prices. Researchers at the Fraunhofer Institute in Freiburg, for example, founded the start-up company Concentrix Solar in 2005, which intends to develop their technology further for the commercial market.

An alternative method to lens concentration of sunlight has been developed by the startup company Prism Solar Technologies (PST) in Stone Ridge, New York. Their photovoltaic module contains a planar concentrator, which distributes the light holographically over the solar cells and in this way concentrates its intensity by a factor of ten. The rows of holographic planar concentrators alternate in the modules with rows of solar cells [3]. PST wants to apply this principle in the future to thin-film technology and thus to reduce costs still more.

However, concentrator cells require direct sunlight; they are therefore suited only for operation in the Sun Belt of the

An example for a large thin-film solar installation based on a-Si is integrated into the roof of the Stillwell Avenue subway terminal in Brooklyn. With a module area of 7060 m^2 (76,000 square feet), this large New York subway station produces a maximum power of 210 kWp (Figure 1). The

INTERNET

PV Status Report 2006 of the European Commission (global overview) re.jrc.ec.europa.eu/pvgis/doc/report/PV_Status_Report_2006.pdf

US Department of Energy
www1.eere.energy.gov/solar

Solar-cell research in the USA (examples)
www.nrel.gov/pv/
photovoltaics.sandia.gov/
www.ece.gatech.edu/research/UCEP/

German Federal Ministry for the Environment...
www.bmu.de/english

Solar-cell research in Germany (examples)
www.ise.fraunhofer.de
www.fz-juelich.de/ipv
www.hmi.de/bereiche/SE
www.iset.uni-kassel.de
www.uni-konstanz.de/photovoltaics

International studies on the long-term behaviour of photovoltaic installations
www.iea-pvps-task2.org

World's largest photovoltaic power plants
www.pvresources.com/en/top50pv.php

Earth. In Spain, Africa, Australia or the southern USA, they could over time become even cheaper than the current conventional cells, even though they require an automatic guidance system to keep them pointing at the Sun. But even with conventional modules, a more adroit utilization of the light can permit increases in efficiency. Bifacial, or "two-faced" solar cells are for example transparent on their back sides at places where they have no contacts. They can thus make use of light which falls onto the back side – e.g. via a mirror. Practically market-ready bifacial cells have already demonstrated an increase of their original efficiencies by up to one-fifth, and laboratory models have shown increases of more than half.

Organic, Polymeric and Dye Solar Cells

Not only "hard" semiconductor materials can convert sunlight into electrical energy. Organic molecules also can have this ability in principle; as a rule, these are dye molecules. Also, some long polymeric-chain molecules act as semiconductors and are suitable in principle for fabricating solar modules. Such organic and polymic solar cells are currently still in the research stage. They are interesting as alternatives to the conventional semiconductor materials since they have some other attractive properties. For example, it is already clear that they can be fabricated cheaply and with a low investment of energy. In contrast to "hard" semiconductor materials like silicon, they do not have high melting temperatures. Furthermore, they can be produced with little strain on the environment. In particular, they offer new possibilities because they are lightweight and flexible. They could, for example, be integrated as dye solar cells into articles of clothing, or within buildings. Solar cells of this type can in principle be printed using ink-jet printer technology.

The organic and polymeric solar cells however still have low efficiencies. Researchers from the Technical University in Ilmenau, Germany, have for example developed laboratory models on the basis of polythiophene, which have efficiencies of 5 %, with an expected improvement in the near future to 7 %. A typical problem with such plastic solar cells lies, to be sure, in the permeability of their molecular networks to smaller molecules; water or oxygen can penetrate them and change their properties, causing their efficiencies to decrease rapdily. The scientists at Ilmenau plan to use encapsulation to slow this ageing process and thus to obtain lifetimes of up to two years.

A further possibility for converting sunlight into electrical energy is offered by the photoelectrochemical cell, also called the Grätzel cell. Michael Grätzel's group at the ETH Lausanne in Switzerland developed this type in the early 1990's [4]. They are also known as dye solar cells or dye solar modules. This type of electrochemical cell, like a battery or an accumulator, contains an electrolyte, currently usually in the form of a gel which is not quite liquid, and two electrodes. The electrolyte in turn contains an organic dye, whose molecules absorb light quanta from the sun-

Fig. 5 *A transparent dye solar module from the Freiburg Institute for Solar Energy Systems, made by silk-screen printing using a special glass technology* (Photo: Fraunhofer ISE).

light and release free electrons. Their interplay with the electrolyte allows the cell to produce electrical energy. The dye can even be made from blackberry or raspberry juice, for which reason the Grätzel cells are well suited for school experiments. The Lausanne group has increased the efficiency of their cells up to about 11 %. In the meantime, their cells have also attained high thermal stability, which is decisive for their practical application: In a long-term experiment of 1000 hours of operation at 80 °C (176 F), the Lausanne cells lost only 6 % of their original power output [5] (Figure 4).

Another direction is being pursued for example by a research group at the Fraunhofer ISE in Freiburg; they are trying to increase the usable area of dye solar modules as much as possible. The problem here is the encapsulation of the electrolyte. The Fraunhofer researchers demonstrated modules with areas of 30 cm × 30 cm in 2006, with which they could fashion graphic elements in different colors, e.g. for advertising signs (Figure 5) [6]. These Freiburg cells have efficiencies of 2.5 %; their developers expect to be able to increase them to 5 % in the coming years.

The developments in Lausanne prove that dye solar modules can offer serious competition to the established technologies in terms of electric power generation on a large scale. However, it will be several years before they reach the photovoltaic market and obtain a significant market share. In the meantime, organic, polymeric and dye solar cells will remain exotic devices which still require more research and development.

Suggestions for Planning a Solar Installation

In the planning of a photovoltaic installation, practical questions are in the foreground. They range from the annual solar radiation at the planned site, the optimal outfitting of the installation and its components, to the question of financing and government subsidies. Whoever wants to set up a small installation on the roof of his or her house should be thinking in terms of a long-term investment. Analyses have shown that for example in Germany, subsidized by a special law called "Erneuerbare-Energien-Gesetz" of the German Federal government in Berlin, a 3 kilowatt installation will amortize itself at the earliest after 18 years. However, the price of electric power in Germany has increased by 40 % between 2000 and 2006. If this trend continues, a photovoltaic installation can become profitable more rapidly in the future.

Naturally, countries of the Earth's sunbelt are the best locations for the use of photovoltaic power generation. But it works also surprisingly well in the higher or lower degrees of latitude, i.e. in countries quite far in the North or South of the Earth. That again shows the example of Germany. In southern Germany, a well-planned and correctly constructed installation with a peak power of one kilowatt – this corresponds to a module area of about 8 m^2 – will have an annual output of more than 900 kilowatt hours. In northern Germany, the corresponding output would be about 800 kWh.

However, a study in the southern German city of Freiburg showed recently that in many photovoltaic installations, construction defects prevented the power yield from reaching its optimum value. Often, the modules are not directed fully towards the Sun or they are in shadow at some times, for example due to trees. Still more difficult to recognise are technical defects in the installation: some solar modules generated too little power, as shown by the study, or the power inverters were poorly matched to the modules or were even defective. If one wishes to operate his or her own installation, it is important to consider such potential sources of problems. In the case of new construction with a solar installation on the roof, the architect and the solar installation constructor should cooperate on the plans from the outset. in order to obtain optimal results.

Summary

The market share of photovoltaics is growing rapidly. But in absolute terms, their contribution to energy generation is still rather small. Offsetting the advantage of a long module lifespan, the required investment costs are high. The still dominant silicon technologies currently require a large amount of energy for cell fabrication in relation to the overall energy which the cells will generate in the course of their lifetimes. New production technologies should change this. In practice, incorrectly installed plants or technical deficiencies often make the resulting efficiency less than optimal.

References

[1] Renewable energy sources in figures - national and international development. German Federal Ministry for the Environment, Nature Conservation and Nuclear Safety, Berlin **2007**. www.bmu.de/english/renewable_energy/downloads/doc/5996.php

[2] PV Status Report 2006 of the European Commission, pp. 51; re.jrc.ec.europa.eu/pvgis/doc/report/PV_Status_Report_2006.pdf

[3] www.prismsolar.com

[4] B. O'Regan, M. Grätzel, Nature **1991**, 353 (24), 737.

[5] CH-Forschung 2002, www.ch-forschung.ch.

[6] Press release of the Fraunhofer ISE on March 30, 2006; www.ise.fhg.de/press-and-media, Press Releases **2006**.

[7] P. Würfel, Physics of Solar Cells. Wiley-VCH, Berlin **2004**.

About the Author

Roland Wengenmayr is the editor of the German physics journal "Physik in unserer Zeit" and a science journalist.

Contact:
*Roland Wengenmayr,
Physik in unserer Zeit,
Konrad-Glatt-Str. 17,
D-65929 Frankfurt am Main, Germany.
Roland@roland-wengenmayr.de*

SOME PEOPLE THINK WE TAKE OUR UNDERSTANDING OF PARTNERSHIP TOO FAR.

When developing our high-efficiency solar cells, we not only rely on our research department, but also on our close cooperation with suppliers and module manufacturers. That's why we go the extra mile to give our partners personal and on-the-spot advice for processing our products and work together to find new solutions for the future.

Q-Cells AG is the world's largest independent manufacturer of solar cells. In 2007, the company is expected to manufacture crystalline solar cells with a total output of 370 megawatt peak (MWp). More than 200 scientists and engineers are working to refine the technology at Q-Cells in order to achieve the aim of cutting the costs of photovoltaics rapidly and permanently, thus making the technology affordable and competitive.

Learn more about Q-CELLS at www.q-cells.com

www.kleinerundbold.com

New Materials for Photovoltaic Energy Conversion

Solar Cells from Ribbon Silicon

BY GISO HAHN

The solar-cell market is booming. But photovoltaic cells are still too expensive to compete effectively with conventional power generation. A notable cost reduction can be achieved if ribbon silicon is used instead of the usual silicon wafers, which are sawed from massive blocks of silicon.

The term *photovoltaic* energy, derived from the Greek, can be translated descriptively as 'electrical energy from light'. In the year 2003, the solar cell, which is at the heart of every photovoltaic module, celebrated its 50th birthday. There are many reasons for the increasing success of solar power. Although in the beginning, satellites were the main users of solar modules as an independent source of electric power, photovoltaic energy soon came down to Earth. In terrestrial applications, a number of advantages play a role: For one thing, photovoltaic power permits a renewable energy supply within a closed cycle, and thus provides freedom from the limited supplies of fossil fuels and their negative effects on the environment. A second advantage is the possibility of a decentralised energy supply for isolated sites with no connection to the power grid, e.g. for mountain cabins, traffic signs or settlements far from power lines. Not least, the modular character of photovoltaic systems is an important reason for the optimistic prognosis regarding the future development of this form of energy conversion.

However, the relatively high energy production costs of power generated by photovoltaic modules have so far put the brakes on the growth of this technology for earthbound applications. While conventional base-demand power plants can generate power today at costs between 1.5 and 3 €-cent/kWh (approx. 2-4 $-cent/kWh), and peak-demand gas turbine plants cost between 8 and 11 €-cent/kWh (approx. 12-16 $-cent/kWh), the generating costs for photovoltaic power currently still fall in the range between 15 and 30 €-cent/kWh (approx. 22-44 $-cent/kWh) [1] for a fully integrated photovoltaic company, depending on the location of installation. In Germany, the "100,000-Roofs Program" and the "Energy Input Law" (Energie-Einspeisungs-Gesetz, EEG) have given photovoltaic power generation a strong boost since 1999: The EEG subsidizes the input of solar power into the power grid at up to 50.5 €-cent/kWh (approx. 73 $-cent/kWh) in the first year of service. Thereafter, the subsidy is reduced by 5% per year. If research and industry succeed in making effective use of the existing potential for savings in the production of photovoltaic solar modules, they may – thanks to these subsidies – soon reach the threshold of production quantities that will yield a further substantial cost reduction due to economies of scale and grid parity will be reached.

The State of the Art

The annual growth rates for installed electric power generated by photovoltaic systems in the past few years were above 30%. The leading nation in sales of photovoltaic modules is Japan, ahead of Europe and the USA (see Figure 1). In the year 2005, new generating capacity of over 1,700 MW was installed worldwide, roughly corresponding to the output of a large nuclear power plant. If growth rates continue at this level, it will thus take several decades more until solar power contributes a substantial portion – 5 to 10% – of the total installed electric power generating capacity. To accelerate this growth, it is necessary to press on with research into various concepts that promise to reduce the cost of solar power as far as possible. The benchmark is the specific generation cost of one watt of photovoltaic power under standard conditions of solar radiation (see infobox "Determining the Efficiency", p. 48); this so called watt-peak cost (W_p cost) has to be reduced.

A bank of furnaces from which silicon ribbons are being pulled from the melt in the form of hollow, octagonal columns up to 6 meters high, using the EFG process.
(Photo: Schott Solar GmbH.)

Renewable Energy. Edited by R. Wengenmayr, Th. Bührke. Copyright © 2008 WILEY-VCH Verlag GmbH & Co. KGaA, Weinheim. ISBN 978-3-527-40804-7

FIG. 1 | THE MARKET FOR PHOTOVOLTAIC MODULES

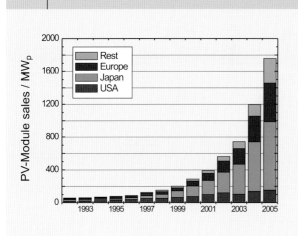
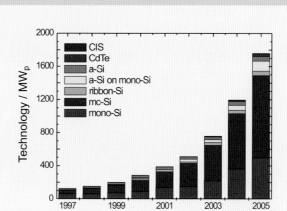

The sales of photovoltaic modules by regions (left) and in terms of the various technologies which are currently relevant (right) [2]. A shift from mono-Si towards multicrystalline Si can be clearly seen.

The first solar cells were fabricated from single crystals of semiconductor materials of the same quality as those used for the production of microelectronics devices. An important factor in determining the suitability of a semiconductor material for photovoltaic applications is the size of its band-gap energy (see infobox "Solar Cells from Crystalline Silicon", p. 46). Gallium arsenide (GaAs) is very suitable, since its band-gap energy of 1.42 eV is perfectly adapted to the solar spectrum. The band gap of silicon (Si), however, is only 1.12 eV, and is therefore somewhat too small. On the other hand, silicon can be produced with the required purity at a much lower cost. Solar power from silicon cells profits here from the years of experience with Si gained in the microelectronics industry. Furthermore, the waste products from the semiconductor industry represent an important source of raw material, since the purity requirements for the fabrication of solar cells are somewhat less stringent than those for microelectronics. For these rea-

sons, GaAs has been used primarily in space applications, where a high efficiency is decisive and production costs are less important. In terrestrial applications, in contrast, reduction of the W_p cost has been the main goal from the very beginning.

For this reason, single-crystal silicon (mono-Si) dominated photovoltaic applications during the 1970s and 1980s. Poly- or multicrystalline silicon (mc-Si) later provided further cost savings. These have to be balanced against the lower conversion efficiencies of cells made from mc-Si, due to crystal defects such as grain boundaries between the individual crystallites, dislocations in the crystal lattice, and the higher impurity concentration as compared to mono-Si. The cheaper starting material however more than outweighs the loss in efficiency. Under suitable process conditions, the W_p costs of mc-Si can be markedly less than those of mono-Si. As a result, mc-Si has in recent years displaced mono-Si from its predominant position and is now the leader, providing 59% of annual installed power [2].

A common feature of both materials is that the flat discs ("wafers") needed for the production of solar cells are as a rule cut out of massive Si blocks ("ingots"). The ingot is first cast; for this process, the highly pure (and therefore expensive) silicon starting material has to be melted (melting point 1,414 °C (2,577 F)) and then allowed to solidify under a well-defined temperature gradient. For an ingot weighing 240 kg (530 lb), this takes about 48 hours. Cutting the individual wafers out of the ingot is accomplished with wire saws, which use wires several kilometers in length and around 200 μm (0.01 in) in diameter. The square wafers in final form are 200 to 300 μm thick and 125–156 mm (5-6 in) on a side. Between 50 and 60% of the starting material is lost in the course of fabricating wafers from the ingots; the major portion of this is literally pulverized in the process of sawing out the wafers. This increases the fraction of the production cost of a solar module due to wafer costs by up to 50% [3] (see Figure 2).

INTERNET

String Ribbon silicon
www.evergreensolar.com/app/en/technology/item/48

Ribbon Growth on Substrate
www.ecn.nl/en/zon/r-d-programme

Ribbon Silicon Technology Movie
www.wiley-vch.de/publish/dt/books/
ISBN 978-3-527-40804-7

Silicon growth techniques
(Silicon programme/Ribbon Growth on Substrate)
www.siliconsultant.com/Pages/SiInfo.htm

Solar energy information
www.solarserver.de/index-e.html

FIG. 2 | MODULE PRODUCTION COSTS

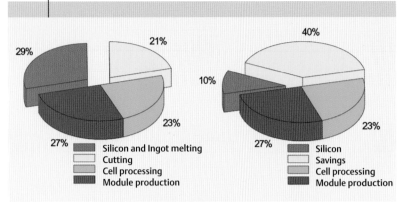

The cost distribution for solar modules. Left: mc-Si wafers sawed from ingots [3]. Right: Wafers of ribbon silicon (RGS) with the same efficiency.

Thin-Film Solar Cells

For these reasons, there has been a search for some time for alternatives to crystalline silicon (c-Si), which is utilized in such a wasteful manner. One such alternative is amorphous silicon (a-Si), which in contrast to c-Si is a direct semiconductor. Direct semiconductors absorb light much more effectively than indirect semiconductors; therefore, the active photovoltaic layer can be made much thinner. In thin-film solar cells, this layer is only a few microns thick and contains only about one percent as much material as in the active layer of c-Si wafers. However, solar cells made from a-Si have the disadvantage that their conversion efficiencies degrade in the first thousand hours of operation. Their efficiency thereafter remains stable, but it is significantly lower than that of c-Si solar cells. This has a negative effect on the W_p costs, since they include other factors besides the production costs of the solar cells; these are area dependent and scale with the efficiency – for example the module costs, cost of the supporting frames, as well as direct costs for obtaining the site where the modules are to be set up. For these reasons, a-Si is applied successfully mainly for devices with low power-output requirements, e.g. for pocket

FIG. 3 | TYPES OF RIBBON SILICON

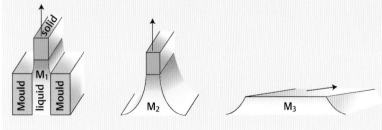

The meniscus at the solid-liquid boundary is used to classify the preparation method for silicon ribbons [4]. M_1: pulling direction vertical, capillary forces push liquid Si up through a mould, after which it solidifies. M_2: pulling direction vertical, the ribbon is pulled from a broad base at the surface of the Si melt. M_3: pulling direction horizontal, the phase boundary extends over a large area.

calculators or watches, where the low production cost is important.

Other materials which are currently the subject of intensive research as absorbers for thin-film solar cells, and are in part already in production, include cadmium telluride (CdTe) and copper indium diselenide (CIS). Although both materials have exhibited high efficiencies in laboratory experiments, the module efficiencies in industrial production are still notably lower than those of c-Si. Many people also consider the use of toxic materials such as cadmium to be problematic, even though it is safely encapsulated in the solar modules.

In spite of intensive research in the area of thin-film solar cells, the market share of photovoltaic modules based on c-Si has increased in the past few years. A main reason for this is the tendency of industrial producers to bank on an established technology in order to minimize risks. It is much simpler for producers to enlarge their existing production capacity than to introduce a completely new process technology. It is thus clear that crystalline silicon will continue to make by far the largest contribution to photovoltaic power generation throughout the coming decade.

Ribbon Silicon

A major portion of the costs of the current c-Si photovoltaic technology is due to the wasteful usage of expensive high-purity silicon. A possibility for reducing wafer costs consists of simply using thinner and therefore more breakable wafers. In principle, it is already possible to reduce the wafer thickness to 100-150 μm and thus to obtain more wafers from a cast ingot and reduce the cost per wafer, if the efficiency and the yield (breakage) remain constant. This is however currently not the case. Furthermore, more "sawdust" would be produced, so that the percentage of wasted silicon would increase.

Another, more elegant method is offered by the use of ribbon silicon. Silicon ribbons are crystalline wafers which are pulled directly from the melt with the required thickness of about 200-300 μm. Ribbon silicon wafers have the great advantage as compared to wafers cut from ingots that almost all the material in the silicon melt can be used to produce the crystalline wafers. The lack of waste "sawdust" gives a significant reduction in the fraction of wafer costs in the overall cost of the modules. In addition, a second cost factor in the conventional process can be eliminated completely, namely the crystallization of the silicon ingots. With the condition that the efficiency of the cells made from ribbon silicon wafers is just as high as that of cells made from wafers sawed out of ingots, up to 40% of the module costs can be saved by using ribbon silicon wafers (see Figure 2). This is the case for so called RGS silicon, which we describe below.

Another important factor is the current, transitory shortage of silicon. Up to now, the photovoltaic industry has been able to fill its needs by using the excess production of silicon for microelectronics. The strong growth in recent times has however had the result that solar-power applications now use just as much silicon as the integrated-circuit producers. Saving silicon material is thus an additional advantage in the use of ribbon silicon, as long as the photovoltaic industry has not yet built up its own silicon production (i.e. in the next few years).

Over the years, many different production technologies for silicon ribbons have been tried out. These technologies can be classified in terms of the meniscus formed by the Si melt at the phase boundary between the liquid and the solid phase (see Figure 3). In the case of meniscus type M_1, the base area of the meniscus is defined by a mould in which the liquid Si climbs up above the surrounding surface due to capillary action. The wafer is pulled upwards out of the melt, at a pulling rate of around 1 to 2 cm/min (0.04 - 0.08 in/min); this is the main factor determining the thickness of the wafer. The heat of crystallization is carried off mainly by radiation, while convection hardly plays a role. Therefore, the rate of crystallization is relatively low and this, in turn, limits the pulling rates.

During the pulling, the temperature gradient at the liquid-solid boundary must be controlled to within 1°C (1.8 F), i.e. from the technological point of view, with a very high precision. This technique was commercialized as early as 1994 under the name 'Edge-defined Film-fed Growth (EFG) and is being further developed by the Schott Solar company in Alzenau (Germany) [5]. The silicon is pulled from the melt in the form of 6 m (20 ft) long tubes. A graphite mould gives it an octagonal shape with very thin walls. This closed shape avoids free edges which would have to be stabilized. A laser then cuts the 12.5 cm (5 in) wide faces of the octagon into square or rectangular wafers.

The meniscus shape of type M_2 has a broader base than type M_1. It occurs when the wafer is pulled vertically upwards directly out of the melt. Owing to the longer meniscus, this type of ribbon silicon preparation can tolerate a greater fluctuation in the temperature gradient at the liquid-solid phase boundary, of around 10 °C (18 F), which permits the use of more compact and less expensive processing equipment. An example of this type is String Ribbon silicon developed by Evergreen Solar Inc. and commercialized since 2001. In this process, two fibers ('strings') made of a material which is kept secret by the producer are passed through the Si melt and pulled parallel, vertically upwards

from the liquid surface (see info box 'Internet' on p. 41). This process makes use of the fact that silicon has a high surface tension, even greater than that of mercury. Thus a silicon film stretches between the two strings, which are about 8 cm (3 in) apart, like a soap-bubble film, and solidifies to a ribbon. A laser then cuts this ribbon into wafers of the desired size. In this likewise vertical pulling method, the pulling velocity is limited for the same physical reasons as in the EFG process to 1-2 cm/min (0.04 - 0.08 in/min).

Considerably higher pulling velocities are possible if the wafer is pulled horizontally instead of vertically from the melt. This is done using the extended meniscus shape of type M_3. For horizontal pulling, a substrate can be used on which the silicon solidifies. This is the case in the Ribbon Growth on Substrate (RGS) silicon process. This method was originated by Bayer AG, and is still in the developmental stage. A belt carrying substrate plates moves under the crucible containing the silicon melt. The silicon is deposited onto the plates. As soon as the wafers have crystallized, they detach themselves from the substrate due to the difference in thermal expansion coefficients, so that the substrate plates are again freed up for the next pass. This process is distinguished by rapid heat dissipation through the substrate plates. Furthermore, the direction of crystallization is decoupled from the horizontal pulling direction; it runs up vertically from the cooler face of the substrate to the upper surface of the wafers. Both these effects permit high pulling velocities up to 10 cm/s (4 in/s), i.e. a much higher throughput than in the vertical methods. This enormously reduces the cost of wafer production.

Crystal Defects and Defect Engineering

All of the types of ribbon silicon wafers discussed here solidify in the form of

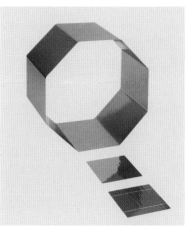

The octagonal tubes obtained from the EFG process are at first cut into sections. Their faces are then cut into rectangular wafers 12.5 cm (5 in) wide. (Graphics: Schott Solar GmbH.)

In the EFG process, the silicon ribbon is pulled out of the silicon melt in the form of hollow, octagonal columns up to 6 meters high, whose walls are only about 200-300 μm thick. (Photo: Schott Solar GmbH.)

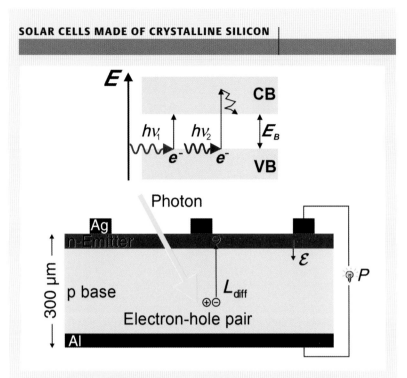

Above: Energy-band diagram. Below: The structure of a silicon solar cell, schematic; Ag means the silver front electrode, Al the aluminium back electrode on the rear side of the wafer.

A solar cell makes use of the internal photoelectric effect, in which incident photons remove electrons out of their bound states. In a semiconducting material, the photon must have at least the energy $h\nu_1$ which is sufficient to excite an electron out of the valence band (VB) over the band gap (of energy E_B) into the conduction band CB (upper part of the figure). In this process, an electron-hole pair is formed.

Photons with an energy $E = h\nu_2 > E_B$ excite electrons into states in the conduction band above the lower band edge. These then give up energy through collision processes until they have dropped back down to the band edge. The excess energy $E - E_B$ is thus lost in the form of heat. The required minimum energy E_B and the loss of excess energy reduce the maximum attainable efficiency of a conventional solar cell made from crystalline silicon down to about 43%. Further loss mechanisms decrease the theoretically achievable efficiency to 28%.

The p-n junction in the solar cell produces an electric field (ε in our diagram). Mobile charge carriers are accelerated by this field in different directions depending on the sign of their charge. This gives rise to charge separation, so that a voltage appears between the emitter and the base electrodes. If the external circuit is closed via a power-consuming device, the electrical power P can be extracted from the solar cell.

In the currently most common type of solar cells fabricated from c-Si, the absorber material must be sufficiently thick that the photons are absorbed as completely as possible in the silicon indirect semiconductor. The absorption is stronger for light of short wavelengths (blue) than for light of long wavelengths (red). In order to use the long-wavelength part of the solar spectrum, and also for reasons of stability, such conventional solar cells are normally made 180 to 300 μm thick. The charge carriers have to diffuse from the site where they are generated to the p-n junction in order to contribute to the current output of the cell. The diffusion length L_{diff} is the distance which the charge carriers pass through before they recombine. L_{diff} is related to the diffusion constant D and the lifetime τ of the charge carriers by the formula

$$L_{diff} = \sqrt{D\tau}$$

In a high-quality solar cell, the charge carriers must have long lifetimes and thus a long diffusion length.

multicrystalline silicon. Their particular process conditions lead to different types of lattice defects. We mention some typical kinds of defects using the examples of the three materials: EFG silicon, String Ribbon silicon, and RGS silicon.

EFG and String Ribbon wafers contain long crystallites of a few cm^2 in area, oriented along the pulling direction, after their production. The dislocation density in the different crystallites varies widely, leading to a very inhomogeneous material quality in the wafers. The main contaminant is carbon, with a concentration up to 10^{18} per cm^3 (10^{25} per cu in); it is thus considerably higher than in wafers which are sawed from Si ingots. Furthermore, metallic impurities may also reduce the wafer quality, however at much lower concentrations.

In the RGS wafers, in contrast, the much higher pulling rates lead to crystallite sizes of less than a millimeter. The dislocation densities, defined as the length of dislocations per unit volume, range up to 10^7 per cm^2 (10^{11} per sq in). By comparison, the wafers used in the microelectronics industry are dislocation-free! The main part of the impurities consists again of carbon, and at somewhat lower concentration, of oxygen.

Impurities and other crystal defects such as dislocations and grain boundaries have a decisive drawback: Because they disturb the translational symmetry of the perfect crystal, they reduce the lifetime of the charge carriers (Figure 4). The defects can form allowed energy levels within the band gap. These 'stepladders' permit an electron which has been excited into the conduction band by a photon to fall back down into the valence band through recombination. It is then lost for further charge transport. The goal is therefore to remove these defect states as far as possible during the fabrication of the solar cells. This increases the charge-carrier lifetimes and improves the quality of the finished solar cell. A longer charge-carrier lifetime for example increases the amount of current which the cell can deliver. More charge carriers reach the p-n junction and can be separated there by the electric field before they can recombine.

One strategy consists of removing defects, for example metallic impurity atoms, during the processing of the solar cells. This 'gettering' makes use of the fact that metals have lower solubilities in silicon than for example in aluminium. Aluminium comes into play in any case, via the back contact of crystalline silicon solar cells: In their industrial production, the back side of the wafers is usually covered with an aluminium-containing paste. This is then burned into the wafer at a temperature of 800–900°C (1,500-1,650 F), forming the rear electrode. At these temperatures, most metal atoms are mobile in silicon; their higher solubilities in aluminium then produce an automatic purification of the silicon wafer, driven by the concentration gradient for diffusion. The metal atoms cause no problems in the aluminium layer, whilst the lifetime of the electrons in the purified, active silicon layer is increased. It is particularly attractive for the producers that this gettering does not require an additional process step, since electrode formation is already part of the production process of solar cells.

Of almost more significance for all solar cells made from mc-Si is the application of hydrogen. Atomic hydrogen can bind to defects such as unsaturated or 'dangling' bonds in the silicon crystal lattice, which would otherwise give rise to defect levels within the band gap. The hydrogen thus modifies the bonds and bonding angles and changes the energetic position of the defect levels. This can result in a shift of the defect levels within the band gap or even their removal from the gap. The art of modifying crystal defects in a positive way is termed 'defect engineering'.

In mass production, a particularly elegant method of applying atomic hydrogen to the wafers has become prevalent in recent years. It makes use of the fact that the front surface of the wafers is covered with an antireflection coating in order to permit the maximum number of photons which arrive at the cell to enter it and produce charge carriers. This is the reason for the characteristic deep blue colour of the crystalline silicon solar cells. This function can be performed by a silicon nitride film, which is deposited from the gas phase using the PECVD method (Plasma-Enhanced Chemical Vapor Deposition). This film binds up to 20 $_{at.}$% of hydrogen during deposition. During forming of the back electrode, the high process temperature causes the hydrogen to diffuse into the silicon substrate, where it can bind to defect structures and shift the defect levels out of the band gap or at least into a more favorable position near the band-gap edge. When the recombination rate due to the defect levels is reduced in this manner, one refers to a ***passivation*** of the defects.

Gettering of impurities and passivation of crystal defects produce a great improvement in the quality of starting material which contains a high defect concentration. This is especially true for ribbon silicon materials, whose quality can vary widely within small regions between neighboring crystallites. The two techniques make it possible to fabricate solar cells with good efficiencies from ribbon silicon wafers (Figure 4). These process steps as a rule do not complicate the production process, since they are already a part of the processing sequence. However, they must be individually optimized for the specific material being used.

In Figure 5, the distribution of current within an RGS and an EFG solar cell is illustrated. While in the EFG material, the stripe-like structures reflect the long crystallites which grow in the pulling direction, one can recognize in the RGS cell in comparison the lower current density and the smaller crystallite size.

Strategies for Cost Reduction

There are two strategies for reducing the cost of power production in photovoltaic devices: Increasing the efficiency of the modules, and lowering the production costs of the starting materials. In the past, research into solar cells was concentrated on increasing their efficiencies. In the meantime, it has become clear that it can make more sense

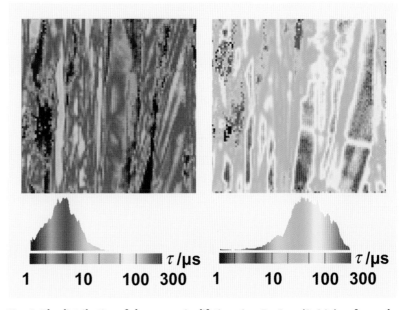

Fig. 4 *The distribution of charge-carrier lifetimes in a 5 × 5 cm (2x2 in) wafer made of vertically-pulled ribbon silicon. Left: In its initial state. Right: After the gettering and hydrogen passivation steps. A lifetime of 10 μs corresponds to a diffusion length of about 170 μm.*

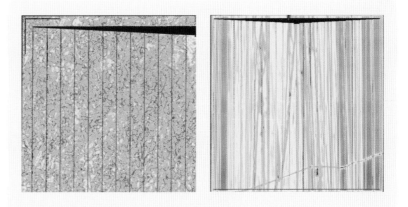

Fig. 5 *The electrical current distribution in solar cells. a) In RGS silicon, with small crystallites (the section shown corresponds to about 1 cm² (0.04 sq in)). b) The current distribution in EFG silicon (4 cm² (0.06 sq in)); the current density increases from blue to green to red.*

economically to produce solar cells from cheaper material of lower quality, whose efficiencies in the end are only marginally lower than those of modules made from more expensive wafers. In such cost-benefit calculations, naturally other considerations based on area costs also play an important role, in which the type of mounting and other factors are taken into account.

A promising candidate for a further notable W_p cost reduction is RGS silicon, because the high throughput of a single wafer production machine and the good use of starting material can lead to the lowest unit costs for mc-Si wafers. However, the small crystallites and the high concentrations of carbon and oxygen are at present limiting factors on the efficiency of solar cells based on RGS silicon wafers. In particular, the charge-carrier lifetimes are still – even after gettering and passivation steps – markedly lower than in other types of silicon ribbons. Current research is therefore concentrating on ways to reduce these impurity concentrations. The initial efforts have already led to an increase in the charge-carrier lifetimes, which gives a corresponding improvement of the efficiencies of solar cells based on this material.

Research has discovered a very elegant method of collecting the major portion of the charge carriers in a material like RGS silicon, with its limited charge-carrier lifetime,

FIG. 6 | CURRENT COLLECTION

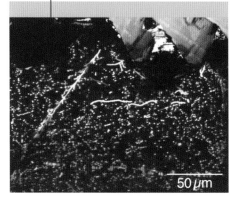

A section through an RGS-silicon solar cell containing a network of n-type conducting dislocations. The network passes through the whole thickness of the wafer and is in contact with the n-conducting electrode at its surface. A bright contrast means a high current-collecting capacity (here made visible by the EBIC technique, Electron Beam Induced Current).

DETERMINING THE EFFICIENCY

The efficiency of a solar cell is found from its current-voltage characteristic curve under illumination. This curve has the same shape as the characteristic of a diode, but is shifted along the current axis by the value of the short-circuit current I_{SC}. In order to take into account the area dependence, usually the current density j is quoted instead of the current I. The open-circuit voltage V_{OC} is the voltage between the electrodes of the cell when no current is flowing.

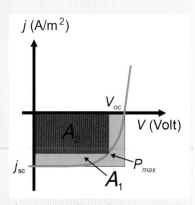

The current-voltage characteristic of a silicon solar cell.

For terrestrial applications, the efficiency is measured under standard conditions. In the AM1.5 standard, the incident light spectrum has a power density P_{in} of 1,000 W/m² (93 W/ft²) and the cell is kept at a temperature of 25 °C (77 F). This corresponds roughly to the average solar radiation at medium latitudes with a clear sky and the Sun at 42° (108 F) above the horizon.

The point of maximum power density P_{max} is reached when the product $j \cdot V$ is maximal. The yellow-green rectangle A_1 (Figure) with sides of length V_{OC} and j_{SC}, is thus larger than the red rectangle A_2 which lies above it and is defined by the point P_{max}. The ratio A_2/A_1 is called the filling factor FF. The efficiency h of a solar cell is defined as the ratio of P_{max} to the power density under light irradiation, P_{in}:

$$\eta = \frac{P_{max}}{P_{ein}} = \frac{U_{oc} \cdot j_{sc} \cdot FF}{P_{ein}}.$$

before they are lost for current output due to recombination [6]. In the p-type base material, oxygen and carbon agglomerates can collect along extended crystal defects such as dislocations. If these agglomerates are sufficiently densely packed along the defects, they form a sort of sheathing around the defect lines. Fixed positive charges along this sheath can then cause a local inversion of the charge-carrier type along the boundary layer, by repelling the majority charge carriers (holes in p-type silicon) and attracting the minority charge carriers (electrons). Thus, an n-type inversion channel is formed along the line of dislocations within the p-type silicon.

Thus, the RGS silicon wafer converts the disadvantage of poor crystal perfection into an advantage: the wafer is penetrated by a branched network of dislocations which reaches up to the surface of the wafer and makes contact there with the n-type emitter layer. The short-lived electrons then no longer have to diffuse to the p-n junction near the surface after being generated by absorption of a photon. It suffices for them to reach the nearest n-type channel, through which they are quickly carried to the emitter layer at the device surface and can contribute there to the external current. Since the average spacing of the dislocation lines in RGS silicon is only a few micrometers high currents can be obtained by this mechanism in spite of the short diffusion lengths (Figure 6). However, the extended space-charge region along the n-type network also causes higher recombination currents in the solar cell (diode saturation currents), so that the improved charge-carrier collection does not automatically lead to a corresponding improvement in overall cell efficiency.

Efficiencies

The efficiency, along with the production cost, is the decisive parameter of a solar cell. It determines the electrical power which the cell can produce. The theoretically attainable efficiency of solar cells based on crystalline silicon under standard conditions of solar irradiation, an AM1.5 spectrum at $p_{in} = 1,000$ W/m² (93 W/ft²), is around 28% (see the infobox "Determining the Efficiency"). The best efficiency which has been obtained in the laboratory for single-crystal silicon is 24-25%, which is close to the theoretical limit. However, the laboratory solar cells fabricated for this measurement are much too expensive for large-scale industrial production. The efficiencies of large-area solar cells which can be mass produced at low cost are currently much lower. For mono-Si, industrial efficiencies in the range of up to 20% can be obtained; for mc-Si, they range up to about 16%.

With solar cells made of ribbon silicon, marked increases in efficiency have been demonstrated in recent years. For laboratory production, the highest efficiencies of String Ribbon and EFG cells are in the range of 18%, and for RGS cells, 14-15% has been demonstrated [7,8]. The highest values obtained, using industrial-scale processing, lie about 1-2% lower (see Figure 7). This dynamic evolution is driven in

FIG. 7 | EFFICIENCIES

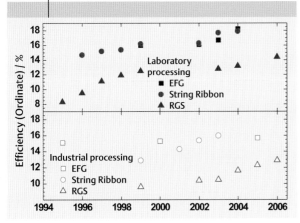

The development with time of the maximum efficiencies of solar cells made of ribbon silicon wafers. The efficiencies of solar cells made by industrial fabrication methods in large-area quantities (below) follow – with a certain time lag – the increase in efficiencies of small-area laboratory cells (above).

particular by the increasing knowledge of the materials properties and the development of processing techniques for solar cells which are adapted to them. New process sequences are first tested on a laboratory scale and then the attempt is made to transfer them to an industrial-scale production environment. In the case of String Ribbon and EFG silicon, efficiencies have in this way already been attained which are comparable to those of solar cells made from mc-Si wafers fabricated by conventional technology. The processes are in both cases of similar complexity and thus have similar costs.

The application of ribbon silicon wafers thus makes a significant reduction in W_p costs possible due to its thrifty use of the expensive silicon starting material. Still greater cost savings should be obtained with solar cells based on RGS silicon wafers. However, their currently low efficiency must be further improved. This requires optimisation of materials quality and of the solar cell fabrication process. If it proves possible to forge ahead to similar efficiencies to those obtainable with the other types of silicon ribbons, the W_p costs could be further markedly reduced. The advantage would then be that producers of crystalline-silicon solar cells could make use of this cost-favorable wafer material in a straightforward way in their existing processing lines, since the main production steps would remain nearly the same. This would greatly facilitate the introduction of solar cells based on ribbon silicon material.

Currently, at the Energy Center of the Netherlands (ECN), an RGS pilot plant is being set up which will deliver RGS wafers on a second by second basis. This production machine is designed to improve the materials quality and therefore the obtainable efficiency, since it works in thermal equilibrium, in contrast to the laboratory devices thus far used. Then, at the latest, ribbon silicon should make a decisive contribution to lowering the production costs of solar modules.

Summary

Photovoltaic power generation will in the foreseeable future mainly make use of crystalline silicon as basic material, with a tendency towards less expensive multicrystalline wafers. At present, the wafers are still sawed out of large silicon ingots, causing considerable material waste by "pulverization". This increases the proportion of the wafer cost in the overall cost of the modules by up to 50%. Ribbon silicon, on the other hand, makes use of a different preparation technology to avoid this material loss, and so yields considerable cost savings. In terms of efficiency, solar cells made from ribbon silicon wafers are already nearly competitive with conventional cells. A further advantage of ribbon silicon is that it can be integrated into existing production lines for solar cells on a crystalline silicon basis. However, the most attractive production processes are not yet mature on an industrial scale.

References

[1] Photon 03 **2007**.

[2] P.D. Maycock, PV News **2006**, *25(4)*, 1.

[3] P.D. Maycock, PV Technology, Performance, and Cost: 1995-2010, PV Energy Systems
Incorporated, Warrington **2002**; obtainable at http://pvenergy.com.

[4] T.F. Ciszek, J. Crystal Growth **1994**, *66*, 655.

[5] F.V. Wald, in: Crystals: Growth, Properties and Applications 5 (Ed.: J. Grabmaier), Springer-Verlag, Berlin **1981**.

[6] G. Hahn *et al.*, Solar Energy Materials and Solar Cells **2002**, *72*, 453.

[7] G. Hahn and A. Schönecker, J. Physics: Condensed Matter **2004**, *16*, R1615.

[8] G. Hahn et al., Proceedings of the 4th World Conference on PV Energy Conversion, Waikoloa, **2006**, 972.

About the Author

Giso Hahn studied physics from 1989 to 1995 at the University of Stuttgart and carried out his diploma thesis at the Max-Planck-Institut for Metals Research there. He received his doctorate in 1999 from the University of Konstanz and was leader of the group for "New Crystalline Silicon Materials" there from 1997 on. He completed his Habilitation at Konstanz in 2005, and has been leader of the Photovoltaics Division since 2006. Since 2007 he has also been employed by the Fraunhofer Institute for Solar Energy Systems in Freiburg.

Contact:
Dr. Giso Hahn,
Photovoltaics Division,
University of Konstanz,
Jacob-Burckhardt-Str. 29, D-78457 Konstanz,
Germany.
giso.hahn@uni-konstanz.de

CIS Thin-film Solar Cells

Photovoltaic Cells on Glass

BY NIKOLAUS MEYER

Currently-used solar modules are made of crystalline silicon: Their fabrication is complex and consumes a large quantity of materials and energy. The Hahn-Meitner Institute in Berlin has been developing thin-film solar cell technology as an alternative. The firm Sulfurcell Solartechnik GmbH, founded in 2001, is making use of this technology to establish a pilot production plant.

The absorption of sunlight is the most important step in photovoltaic energy conversion: only light which has been absorbed can be transformed into electrical energy. Silicon is a semiconductor; it absorbs light only when the energy of the incident light quantum (photon) is equal to or greater than the band gap energy of the material. In that case, the energy of the photon induces an electron to jump out of the valence band into the conduction band, across the "forbidden zone" or band gap. Crystalline silicon is however an indirect semiconductor, so that for this transition to take place, the momentum of the electron must also be changed by interaction with the thermally vibrating crystal lattice. Since a simultaneous change in the energy level and the momentum of an electron is less probable, it requires a disc of crystalline silicon (wafer) of 0.2 mm (0.01 in) thickness in order to completely absorb the sunlight.

CIS – a Direct Semiconductor

For photovoltaic applications, direct semiconductors are therefore more interesting; their valence electrons do not have to change their momenta in order to accept energy from a photon. This class of semiconductors includes the chalcopyrite compound copper indium disulfide ($CuInS_2$, abbr.: CIS), a semiconducting compound of copper, indium und sulfur. Since CIS can completely absorb incident sunlight even as a very thin film of about 1 Ìm thickness, the amount of semiconductor material required to make a solar cell from it is only about one hundredth of what is required by conventional silicon technology. Furthermore, the band gap of CIS, amounting to 1.5 eV, corresponds more closely to the

A small module of CIS thin-film solar cells (Photo: Hahn-Meitner-Institut).

solar spectrum, so that a maximal proportion of the sunlight can contribute to photovoltaic energy conversion. The absorption coefficient and the band gap thus make CIS an ideal absorber material for the fabrication of solar cells. Once the negatively-charged electrons have been excited into the conduction band through light absorption, a corresponding positive charge is left in the valence band in the form of quasi-particles, called "holes". The second step in photovoltaic energy conversion now consists of separating the (excited) electrons from the holes. This task is performed by a p-n junction. It is fabricated by connecting the p-type conducting absorber to a second semiconductor layer which has n-type conduction. In a p-n junction, an electric field develops, and it separates the electrons from the holes. In the case of a CIS solar cell, the n-type semiconductor zinc oxide (ZnO) is employed as the n-conducting layer of the p-n junction. As in the negative pole of a battery, the negative charges collect in the zinc oxide layer when the cell is exposed to light, and the positive charge carriers collect in the (p-type) absorber. In order to use this stored energy, the charge carriers must enter and move through an external circuit. To allow this, the p-n junction is fitted with electrical contacts or electrodes – in the case of the CIS/ZnO junction, this is accomplished by a back contact of molybdenum metal and a front contact of highly conducting zinc oxide (n-ZnO). In the language of semiconductor physics, the structure Mo/CIS/ZnO/n-ZnO completely describes the CIS solar cell. In practice, it has however been found that good-quality solar cells are obtained only if a very thin buffer layer is inserted into the contact region between the CIS and the ZnO layers (Figure 1). Exactly what is the role played by this buffer layer is, by the way, still not clear. In contrast to silicon, which is one of the best understood materials known, CIS is a "young" material: Some of its basic properties are still not described or understood. Technologically, however, the CIS solar cell is already interesting today, even though its potential has probably not been explored to anywhere near its full extent.

Glass Coating in Place of Wafer Technology

Glass is the substrate material of a CIS solar cell, and coating the glass substrate makes this otherwise passive construction material into a solar module for generating electric current. Fabricating CIS solar cells from wafers is neither possible nor reasonable – this can already be seen by the fact that the thickness of the photovoltaic active layer is of the order of 1 μm and is thus very small. Instead, glass

Renewable Energy. Edited by R. Wengenmayr, Th. Bührke. Copyright © 2008 WILEY-VCH Verlag GmbH & Co. KGaA, Weinheim. ISBN 978-3-527-40804-7

serves as a low-cost substrate material, and the other materials which make up the solar cell are deposited layer-by-layer onto the glass surface. All together, the polycrystalline layered structure has a thickness of about 3 μm – CIS technology is thus a thin-film process.

For scientific purposes, thin films are often produced by thermally evaporating the desired material and allowing it to condense onto the substrate. For industrial processes, this is not a favorable method, since the apparatus must work at high temperatures – copper evaporates only at temperatures above 1,300 °C (2,400 F), for example – and the deposition rates are not so easy to control. For glass coating, cathode atomization or "sputtering" has therefore become the method of choice: upon bombardment with accelerated ions, atoms or clusters of atoms are dislodged from the surface of a sample and precipitate onto the surroundings. For this technique, the desired sample material in the form of a solid block is introduced into a high-vacuum system, where it serves as the cathode for an electric field into which a noble gas (usually argon) is passed. If the field strength is sufficient, the gas is ionized by the field, a plasma ignites and the ionized gas atoms are accelerated towards the cathode. Their impacts on the sample "atomize" the cathode material (thus "sputtering"), and as in thermal evaporation, it enters the gas phase. This atomized material condenses onto the cool glass surface opposite the cathode. The plasma and the sample-substrate geometry can be adjusted so that uniform layers of high electronic and optical quality are produced, which can cover areas of up to several square meters (1 m² ≈ 10 sq ft). Sputtering technology is used to apply the molybdenum and zinc oxide layers of a CIS solar cell; sputtering of the CIS compound itself has not yet been achieved. Nevertheless, in order to exploit the advantages of sputtering, the Hahn-Meitner Institute has developed a two-stage fabrication process (Figure 2): First, precursor layers of copper and indium are deposited – for this purpose, sputtering can be employed. This step determines the thickness of the layer and the composition of the final CIS film, and the use of sputtering avoids the formation of inhomogeneous layers. To form the CIS compound, the precursor layers are heated to 500 °C under an atmosphere of sulfur vapor. The high reactivity of the sulfur leads to the formation of the compound copper indium disulfide (CIS) within a few minutes during this process step.

A single CIS solar cell can currently generate a power output of about 13 mW/cm² (84 mW/sq in) under irradiation by the midday Sun. A solar cell of 1 m² (10.8 sq ft) area would produce a current of up to 200 A at a voltage of 0.65 V. This high current would increase the power losses in the current-carrying elements of the solar cell. For this reason, for a large-area photovoltaic element, the overall area is divided into small individual solar cells which are connected in series to form a complete module. In contrast to the case of silicon technology, this series circuitry can be integrated into the fabrication process (Figure 2): After de-

FIG. 1 | A CIS THIN-FILM SOLAR CELL

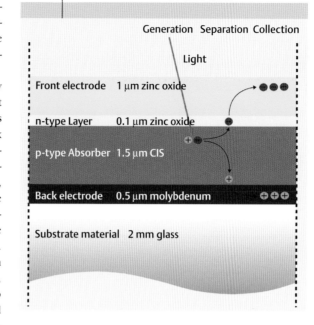

The structure and function of a CIS thin-film solar cell in layers (schematic).

position of the back electrode, after deposition of the buffer layer, and after the preparation of the front electrode, slightly shifted lines are scribed into the layers. This yields strip-shaped solar cells of 5 to 10 mm (0.2-0.4 in) width, which are connected via a narrow contact from their front electrodes to the back electrodes of the neighboring cells, and which therefore work like batteries connected in series.

On the Threshold of Commercial Success

In the laboratories of the Hahn-Meitner Institute, CIS modules of 5 × 5 cm (2 × 2 in) area have efficiencies of 10 %, i.e. they convert one tenth of the incident sunlight into electrical energy [1]. The theoretical limit of 25 % has not yet been attained. The module efficiency however approaches that of commercial silicon modules, which lies between 13 and 15 %. The CIS technology is industrially interesting due to its materials and process advantages: 99 % lower requirements for the expensive semiconductor material, about one-third fewer process steps, and energy consumption reduced by two-thirds more than compensate for the lower efficiencies as compared to conventional silicon modules. For these reasons, market observers predict an important market share for the CIS technology.

Encouraged by these advantages, scientists from the Hahn-Meitner Institute, together with a consortium of private and public investors, launched the enterprise Sulfurcell; in May of 2003, it began applying the CIS technology to large-area modules and to setting up a pilot plant for device fabrication. The industrial development work remaining to be done still involves challenging tasks: a large-area, rapid heating process must be developed, a stable production process must be established, and a reliable method of encapsulating the modules found. The production of the thin films is only one part of this overall module fabrication process; long-term protection of the active thin films against

FIG. 2 | THE COATING PROCESS

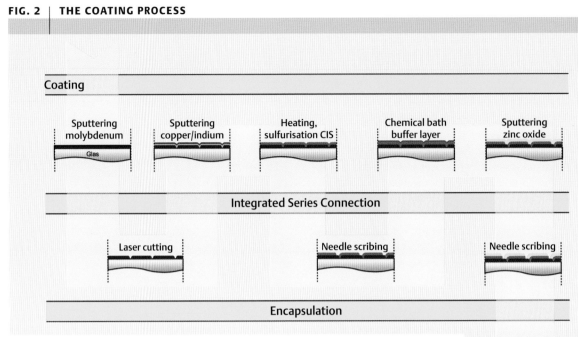

Coating

| Sputtering molybdenum | Sputtering copper/indium | Heating, sulfurisation CIS | Chemical bath buffer layer | Sputtering zinc oxide |

Glas

Integrated Series Connection

| Laser cutting | Needle scribing | Needle scribing |

Encapsulation

Lamination

Coating processes for the fabrication of CIS modules. First, glass substrate plates are coated with metal and semiconductor films via cathode sputtering. Then, in a tempering step, precursor films of copper and indium are converted to the compound semiconductor CIS under sulfur vapor. In between the coating steps (above), the surface is structured with a laser beam or by mechanical scribing: this produces strip-shaped solar cells which are connected in series into a module (center). Multiple coating and scribing guarantees the correct contacting of the current-carrying layers to give a lateral series circuit of the individual cells (here from the right to the neighboring cell on the left). After deposition of the front contact of zinc oxide (yellow), a second glass plate is laminated onto the structure in order to protect the active layers from the atmosphere (encapsulation) (Graphics: Sulfurcell).

atmospheric degradation is another. The path to solutions of these problems will not always be straightforward. It will require some time before new materials will be accepted for industrial applications. But the CIS technology offers a great opportunity to make photovoltaics more economically feasible. For this reason, Sulfurcell and its partners have taken the decisive step: In 2006, the first CIS modules were placed on the market.

Researchers become Entrepreneurs

There has been no lack of visions for the founders of the technology firm Sulfurcell: to revolutionize the solar-energy market using thin-film technology; to market solar modules as low-cost construction materials; and to make photovoltaic electrical energy competitive on the power market. But there was initially no interest from the industry when the Hahn-Meitner Institute (HMI) first developed its technology in 1999. The general economic situation was too weak and the path from the laboratory to the production plant appeared too long. In order to apply the technology in spite of this lack of interest, an entrepreneurial initiative had to come from within the Institute. At the suggestion of the division leader, Martha Lux-Steiner, a team of HMI employees began to prepare the founding of a commercial enterprise in 1999, which was planned to set up a pilot production plant for CIS modules. Over three years later, the Sulfurcell firm had been founded and its financing of over 15.6 M € had been secured. Our team, which prepared the way for the startup of Sulfurcell, had to rethink its plans and launch new initiatives several times during the various stages of the financing. The presentation foils with which we first began soliciting support for the founding of Sulfurcell in the Fall of 1999 were characterized by physical descriptions. An entrepreneurial consultant confronted

us back then for the first time with the question of financially effective cost advantages. We could not supply such important economic performance figures, neither for our own technology nor for the competition.

But investors and bankers are not semiconductor specialists; they invest in business plans and entrepreneurs. So we had to translate the physics into business economics. In the process, we learned to deal with problems which are unfamiliar to most scientists: What effect will having one less dopant element have on the production costs? How will faster processing times affect the personnel requirements? What cost advantage is to be gained by using sulfur instead of selenium? Such questions cannot be answered precisely on a scientific basis, but we could obtain empirical values and realistic cost estimates. We invested a half year in analyzing the planned production of CIS modules economically and in acquiring an overview of the market. This put us in the position of being able to communicate the goals of our enterprise concisely and clearly: Sulfurcell intends to produce solar modules for 50 % less than those currently on the market. The economic potential of the CIS technology finally convinced firms in the glass sector and in mechanical construction, middle-scale entrepreneurs and finance investors. In terms of the financing, however, this was only half the task – for the remainder of the path, it was decisive how the risks of our project would be evaluated. Even without an understanding of the details of the production process, it was clear to every potential investor that Sulfurcell, like any other enterprise with a completely new technology, is a promising but risky financial investment. We were after all planning on at least three years without any business turnover and three more years until the firm could show a net profit.

This is a long period of time, since apart from any possible technical complications, markets can collapse, new competition can arise, or key persons with important know-how can drop out. Banks are not candidates for the financing of such an enterprise; venture capitalists have become more reticent since the collapse of the New Economy, and many industrial organisations have begun to concentrate on their core businesses in view of the economic stagnation. In order to nevertheless procure financing, we had to systematically reduce the risks of the project. Sulfurcell has secured suitable partners for supporting the establishment of the enterprise and has built up strategic co-operations. This was not possible without service in return: a stake in the profits and in the firm had to be granted, sales channels mapped out in advance, and know-how disclosed. The chances of profit and the risks had to be divided among the founders and the partners in a transparent manner. If the business plan and the financial concept are convincing, then in the end, personal impressions and preferences are decisive as to whether a financial commitment will be made. One financial investor later revealed that his fundamental decision to take a holding in the project was already clear on his way home from the first presentation. Nevertheless, there followed a comprehensive examination of the business plan before an official offer of participation was made. We met with a middle-scale entrepreneur from Southern Germany in a relaxed atmosphere with pretzels and beer and presented our plans to him. A few days later, he offered via e-mail to invest a large sum. But his reservations towards the other partners in the end prevented his commitment: Mutual trust among the partners is an indispensable factor in the financing of a new enterprise. Finally, a network composed of affiliates of the HMI, public funding sources, industrial investors and venture-capital providers made it possible for us to arrive at our goal and obtain financing for the pilot-plant production. Among the shareholders in Sulfurcell are currently the specialist for building services M+W Zander (Jenoptik AG), the investment branch of the power generating firm Vattenfall,

and the Berliner Energy Environment Fund, which is sponsored by the Bewag and Gaz de France. One thing is certain: there is no *single* recipe for success for the founders of a start-up firm. Just as personalities and ideas are different, so are the ways of starting up one's own enterprise. In the end, the decisive factor is a determination to create something new, in spite of all the imponderabilities, and to make a success of it.

Summary

Present-day solar modules are made of crystalline silicon. Their fabrication demands an elaborate process technology and consumes a large quantity of materials and energy. The Hahn-Meitner Institute (HMI) in Berlin has been developing thin-film solar cell technology as an alternative. In this technology, copper indium disulfide (CIS), a direct semiconductor, is employed as absorber. It can convert the incident photons of sunlight directly into free charge carriers and thence into electric current. Thus, CIS solar modules require only a very thin semiconductor film, about 1 μm thick, which is deposited onto a glass substrate. CIS modules are estimated to be about 50 % cheaper to produce than current silicon solar cells. The Sulfurcell Solartechnik GmbH enterprise, founded on the basis of the HMI technology, is now setting up for pilot-plant production.

References

[1] J. Klaer *et al.*, Thin Solid Films **2003**, 431, 534.

About the Author

Nikolaus Meyer, together with Ilka Luck, founded the firm Sulfurcell Solartechnik GmbH in 2001 as a spin-off from the Hahn-Meitner Institute (HMI), and he is currently its managing director. He studied physics at the University of Hamburg and the TU Berlin, and managerial economics at the Hagen University. He obtained his doctorate at the FU Berlin for work on thin-film modules performed at the HMI, and worked at Siemens AG on their industrial fabrication. The founding of Sulfurcell was awarded a prize in 2001 by the Business Plan Competition of Berlin-Brandenburg.

Contact:
*Dr. Nikolaus Meyer,
Sulfurcell Solartechnik GmbH,
Barbara-McClintock-Str. 11,
D-12489 Berlin, Germany.
info@sulfurcell.de*

INTERNET

Homepage of Sulfurcell Solartechnik
www.sulfurcell.de (click on "English")

Geothermal Power Generation
Energy from the Depths of the Earth

BY ERNST HUENGES

The Earth can provide enough heat energy to drive geother-
mal base-load power plants – everywhere, not just in active
volcano areas. However, tapping this energy requires drilling
wells down to depths of several kilometers.

Ambitious energy and environmental policy goals are creating new challenges for energy suppliers: The energy mix of the future will have to be ecologically friendly, secure in resources, competitive, and most especially sustainable. This will require a reduction of emissions and a clear-cut increase in the proportion of renewable energies for the supply of electric power.

The goal of the European Union to at least double the proportion of renewable energy sources in the overall energy consumed in the EU by the year 2010 makes it clear just how high the expectations vis á vis the energy producers are. Geothermal energy is becoming more and more interesting in the course of this process: In contrast to wind and solar energy, it is available around the clock, which makes it especially attractive for base-load power plants. It represents an ecologically exemplary and expedient alternative to nuclear power and to fossil fuels.

Energy in the form of technically usable heat or electric power can be produced from geothermal sources as needed. The Earth contains a high potential for supplying heat to the energy economy. Its heat content results from the release of gravitational energy by the contraction of gas and solid particles during its formation, as well as from primordial heat dating from the early days of the Solar System, and from the energy released due to the decay of radioactive isotopes. According to current knowledge, the isotopes which are significant for heat production are those of uranium, thorium and potassium that are enriched in the continental crust, which consists mainly of granitic and basaltic rock (see the infobox "Heat from within the Earth" on p. 57). The occurrence of geothermal heat sources is therefore not limited to regions with noticeable volcanism. In principle, there is geothermal heat everywhere, including under Central Europe. But here, one must drill down to depths of four to five kilometers (about three miles) in order to tap a level of temperatures which is high enough to effectively generate electric power using steam turbines. This potential can be utilized only after the costs and risks of its development have been effectively reduced. The challenge lies in the establishment of technologies which improve the

The Geothermal Labora-
tory in Groß Schönebeck

yield of geothermal repositories and reduce the risks associated with their exploration and exploitation.

Geothermal Energy is Still Exotic

On a worldwide scale, geothermal energy is still an exotic source. Only 0.35 % of the globally produced thermal energy is derived from geothermal sources, according to UN statistics. Of the installed electrical generating capacity, all the geothermal plants together produced not quite 8 GW in the year 2000 [1]. This corresponds to about 1.6 % of the world's power production from renewable sources, which is dominated by hydroelectric power [2]. For comparison: In 2003, the total installed power output of all the world's wind generators was nearly 40 GW [3]. However, wind power depends on the weather, so that all the wind power plants on the globe practically never output their maximum power at the same time. In the case of geothermal power, in contrast, this would be possible, at least in principle. Unlike Iceland, in Central Europe geothermal sources play only a subordinate role. Only Italy can claim a significant number of geothermal power plants, which at least produce more than 800 MW of electric power. Lardarello is also the birthplace of the extraction of electrical energy from geothermal heat: In 1904, Count Piero Ginori Conti installed a dynamo there which was driven by steam from the volcanic terrain. It lit up five incandescent lamps in the village. In Germany, the utilization of geothermal energy has seen relatively high growth rates in recent years. At the end of

Renewable Energy. Edited by R. Wengenmayr, Th. Bührke. Copyright © 2008 WILEY-VCH Verlag GmbH & Co. KGaA, Weinheim. ISBN 978-3-527-40804-7

2005, 570 MW$_{th}$ of heat power from geothermal sources was installed. Of this, 65 MW$_{th}$ was supplied to larger plants, while a further 505 MW$_{th}$ was obtained from geothermal probes: these are heat sources for heat pumps which typically are used to heat single- or multiple-family dwellings. Thus, all together 0.8 % of the overall heat requirements in Germany were supplied from geothermal sources, which is still a very small amount. In the Fall of 2003, the first geothermal power plant in Germany went online in Neustadt-Glewe (Mecklenburg), with an electrical output power of about 200 kW.

Geothermal Energy Sources

Heat can be obtained from the Earth in various ways. Relatively widespread in Central Europe are near-surface geothermal sources: Heat pumps use surface and ground water from a depth of a few meters as a heat source for space heating in houses. Installations of this type require only a few

degrees of temperature difference in order to provide sufficient heating. A second heat source is hot water from deeper within the Earth. Such hydrothermal systems can be found in areas with active volcanism, but also in non-volcanic regions. Today, most of the large geothermal power plants in the world use hot water from volcanically active regions to generate electric power. A typical example of the advantages and disadvantages of this type of energy source is the 60 MW power plant at the Krafla volcano in Iceland: It delivers hot hydrothermal brine at initially almost 400 °C (750 F) from about 20 wells of up to 2,200 m (7,200 ft) depth. The water expands and cools on rising up the well shaft and finally is input to the power plant as steam at 170 °C (340 F). Since these brines contain a large amount of carbonic acid, hydrogen sulfide, salts, and heavy metals, they are corrosive and toxic. This water threatens to attack the well casings and produces waste water which burdens the environment, insofar as it is not pumped back

FIG. 1 | GEOTHERMAL POWER GENERATION

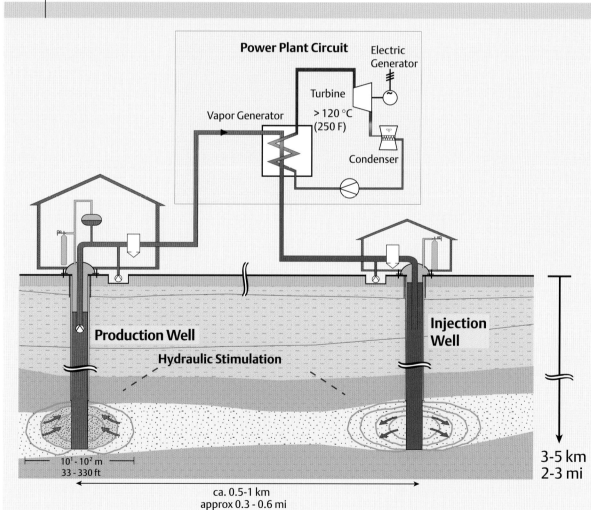

A pump brings hot water through a production well from deep within the Earth to the surface. Its heat is used in a vapor generator to drive a turbine for electric power generation within a power plant circuit. The turbine circuit contains an organic working fluid with a low boiling point in order to increase its efficiency (Organic Rankine Cycle). The cooled water is pumped back to the depths through an injection well (blue). (Graphics: GFZ.)

into the depths (this is currently under test at Krafla). Owing to the composition of the hydrothermal brines, such power plants emit hydrogen sulfide and carbon dioxide – the latter however in notably smaller amounts than from the combustion of fossil fuels.

Hydrothermal systems which are not directly connected with a volcano cause many fewer problems. In Southern Germany and on the North German Plain, there are for example several regions with hydrothermal low-pressure repositories at depths up to about 3.000 meters (10.000 ft). These areas of hydrothermal potential are aquifers which carry hot water – usually salty – that can be brought to the surface through wells.

Since this water is at temperatures between 60 and 120 °C (140-250 F), it is hardly suitable for effective electric-power generation. Therefore, it is mainly used for space heating. The heat from the deep-well water is transferred in heat exchangers to the district heating grid. In this range of temperatures, there are a number of possibilities for utilizing the geothermal heat – apart from electric power generation. Typical examples are central heating installations for local and district household heating, small consumers and industrial applications. The direct use of thermal water for bathing and in therapeutic baths is a classic example.

Hot and Deep

Below a depth of 4,000 meters (13,000 ft), one finds practically everywhere under the Earth's surface rock formations with temperatures above 150 °C (300 F). They contain by far the largest reservoir of geothermal energy which is currently technically accessible. Hydrothermal systems or Hot-Dry-Rock systems (HDR systems), as these formations are called, depending on their water content, thus represent a great future potential for geothermal applications.

'Dry' here means that not sufficient natural water is present in order to pump it over a long period of time to the surface, as in the hydrothermal repositories. A geothermal HDR power plant therefore must pump the water back to the depth and force it through the hot rock. After it has made its way through this natural heat exchanger, it can be pumped back to the surface to again provide its heat energy.

FIG. 2 | **THE GROSS-SCHÖNEBECK BOREHOLE**

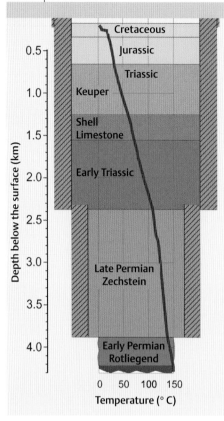

Lithological classification and temperature profile in the Groß Schönebeck borehole. For geothermal electric power generation, the Early Permian Layer is particularly interesting, since it contains hot water at depth with the necessary temperature.

Hydrothermal power plants, in contrast, make use of naturally present water from the depths. This requires the drilling of injection and production wells (Figure 1). Often, the rock formations at the bottom of the wells have natural fissures which are too small; the resulting water throughput is too low and the effective surface area is not sufficient for useful heat exchange. Using special fissure stimulation methods, artificial fissures must then be produced and the existing ones enlarged.

One such method is hydraulic stimulation of the cleavages and fissures. Hydraulic fracturing is a well-established procedure in the petroleum and natural-gas industry. Developed in the 1940's and continuously improved, it is employed there to increase the productivity of oil and gas wells. Hydraulic fracturing is increasingly playing a key role in the exploitation of geothermal heat, also. In this way, the natural water throughput of the reservoir rock can be increased through active stimulation to the point that geothermal energy production becomes economically interesting. However, the stimulation methods used in petroleum exploration are applicable only to a limited extent in the geothermal exploitation of hot water resources. They must be further developed and matched to the needs of geothermal well drilling.

The development of suitable technologies for the exploitation of underground heat has been one of the central research goals at the GeoResearch Center in Potsdam (GFZ), Germany, in the past years. In Europe, this development is also being carried out at Soultz-sous-Forêts (Alsace, France) for application to HDR processes. These research and development projects combine inter-disciplinary basic research for the characterization of potential geothermal repositories with economic and technological planning concerning the operation of geothermal installations.

The scientists at the GFZ can in particular apply their expertise to the investigation of the geological, geochemical, geophysical, and geomechanical aspects of geothermal site development. In addition, the analysis and evaluation of the overall systems is carried out.

For hydraulic experiments and borehole measurements, the GFZ has an *in situ* research laboratory at a 4.3 km (2.7 mi) deep well in Northeast Brandenburg (Northeastern Germany. In November 2003, in a large-scale experiment, the method of massive "waterfrac" was tested for the first time on 150 °C (300 F) deep sedimentary rock layers in the North German Basin. With success: Following this stimulation operations, the productivity of the well increased for the first time into a range which not only generally permits geothermal power generation in the North German Basin, but also makes it interesting from the economic point of view.

The Geothermal Laboratory Groß Schönebeck

These extensive experiments, which I shall describe in the following, could be carried out only through teamwork combining the project group *Geothermie* from the GFZ Potsdam with the work of other experts: researchers from the Federal Institute for Geosciences and Natural Resources in Hannover and from the Technical University of Berlin, water experts from the Neubrandenburg firms GTN GmbH and Boden Wasser Gesundheit GbR, as well as specialists from the Bochum firm MeSy GEO-Messsysteme GmbH all participated.

Groß Schönebeck was chosen as the site of the experimental borehole by the GFZ on the basis of geological and technical data analyses. Since drilling a well of some kilometers in depth is rather expensive, only previously existing old wells were taken into consideration; opening them up again by boring out the cement filling costs considerably less than drilling a completely new well. In addition, old wells have the advantage that their drilling records already contain detailed information about the underground strata. These records include information about the rock formations encountered, properties of the rock, cementation transcripts, drilling reports and other important data.

The scientists of the GFZ researched the drilling records of more than fifty old wells which initially appeared to be suitable. They finally chose the natural-gas exploration well E GrSk 3/90 in Groß Schönebeck, drilled in 1990. On reopening, the well was deepened by 54 meters (148 ft) to a depth of 4,294 meters (14,088 ft). Thereafter, the borehole was available as an *in situ* experimental and measurement laboratory for carrying out borehole measurements and experiments.

It accesses geothermally interesting strata in the North German Basin at depths between 3,900 und 4,300 meters (13,000 and 14,000 ft) and at temperatures around 150 °C (Figure 2).

Down to a depth of 3,873 meters (12,707 ft), it is cased telescopically. At the surface, it has a diameter of about 24.5 cm (9.7 in); at its deepest point, of 12.7 cm (5 in). In October 2003, the borehole was deepened somewhat further, so that its final depth is now 4,309 m (14,137 ft).

Before the fissure stimulation of the rock layers, it was first important to document the initial state of the borehole. To this end, in 2001 hydraulic tests and borehole measurements were carried out. Furthermore, rock samples had

HEAT FROM WITHIN THE EARTH

The Sun, in fact, radiates 20,000 times more energy onto the Earth than reaches its surface in the form of heat from the depths. Nevertheless, geothermal heat is a practically inexhaustible source of energy on a human scale. This heat energy has three origins:
- The gravitational energy stored in the interior of the Earth;
- the primeval heat energy stored in the Earth's interior; and
- the decay of natural radioactive isotopes.

As the Earth was formed from the protoplanetary nebula by accretion of matter, i.e. chunks of stone, dust, and gases, its mass increased and with it its gravitational field. Thus, the matter that continued to rain down on the nascent Earth impacted with increasing force, and the gravitational energy released was converted for the most part into heat. Although a large portion of this heat was radiated back into space estimates show that an energy between 15 and 35×10^{30} J remained in the proto-Earth. An additional amount of energy came from the heat which the matter from the protoplanetary nebula brought to the nascent Earth.

In the Earth's continental crust, the decay of natural radioactive isotopes makes an important contribution to the geothermal heat. Especially in the layers near the Earth's surface, the naturally-occurring isotopes ^{40}K, ^{232}Th, ^{235}U, ^{238}U und others are enriched in the granitic and basaltic rocks. In the basaltic rocks, the radiogenic heat production leads to a power output of around $0.5\ \mu W/m^3$ ($0.01\ \mu W/cu$ ft); in granites up to $2.5\ \mu W/m^3$ ($0.07\ \mu W/cu$ ft). Since the formation of the Earth, this source of heat has released at least an estimated 7×10^{30} J of energy. According to modern estimates, these three sources lead to a total heat energy stored in the Earth of between 12 and 24×10^{30} J. The Earth's outer crust down to a depth of 10 km (6.2 mi) thus contains about 10^{26} J. The resulting heat current towards the surface is about 65 mW/m^2 (6.04 mW/sq ft). For each kilometer downwards into the Earth, the temperature in the outer crust increases by 30 Kelvin on average.   *RW*

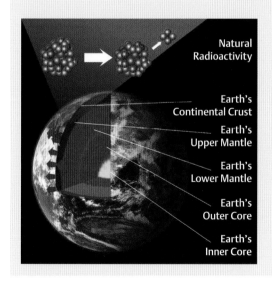

Natural Radioactivity

Earth's Continental Crust

Earth's Upper Mantle

Earth's Lower Mantle

Earth's Outer Core

Earth's Inner Core

In the continental crust, natural radio-activity produces strong geothermal heat (Graphics: Roland Wengenmayr).

to be characterised by laboratory studies and measurements in the borehole.

Stimulation Increases Productivity

The economically-competitive conversion of geothermal heat into electrical energy requires temperatures above 150 °C (300 F), which in large areas of the North German sediment basin can be found at depths between 4,000 and 5,000 meters (13,000-16,500 ft). Alongside these minimum temperatures – as already mentioned – a stable production of large amounts of thermal water is a further precondition to successful energy extraction.

FIG. 3 | STIMULATION

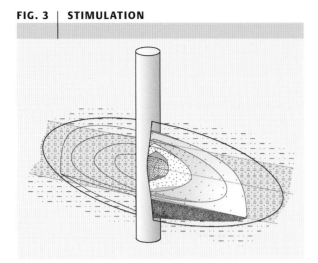

The drawing shows how stimulation produces a vertical crack in the borehole. An artificial fissure follows the stress field at depth: it grows in a direction parallel to the principal stress axis.

To fulfill these conditions, the rock strata must be highly porous and readily permeable, i.e. they must have a large concentration of hydraulically connected pores (hot fractured rock). This permits a good flow of water through the rock and a high throughput to the wells. However, at depths with minimum temperatures above 150 °C (302 F), the natural permeability of the rock is low. It must therefore be artificially broken up by fissure stimulation, in order to allow an unimpeded circulation of the water. Along with the production of a widespread fissure system, a stimulation experiment such as ours must also establish a connection to water-bearing cracks which are naturally present (Figure 3).

To accomplish this, a fluid is pumped into a borehole in a short time and at high pressure. This fluid is usually water, as it was in our case. During the stimulation, the pressure of the fluid exceeds the stress which is present in the rock layers; thus it can enlarge fissures already present in the rock, interconnect them, and produce new cracks (hydraulic fracturing). The injection rates are increased stepwise, and the fluid may be mixed with highly viscous additives.

If necessary, it is also mixed with a fissure propping agent to secure the opened fissures – for example ceramic balls of ca. 1 mm (0.04 in) diameter. These collect in the hydraulically-produced fissures and hold them open after the pressure is released. The stimulation thus produces a widely branched system of cracks which gives the thermal water new flow paths to the production well. It functions both as a transport route and as a below-ground heat exchanger with a large contact area.

Stimulation of the Sandstone

The first – relatively gentle – stimulation experiments in the sandstone strata at 4,200 meters (13,800 ft) depth were carried out at Groß Schönebeck in a largely conventional manner, that is on the basis of experience from petroleum and gas exploration. We injected several hundred cubic meters of a highly viscous fluid, a special gel, at an overpressure of 17 MPa (2.3 Gtorr), and introduced fissure propping agents. In fact, measurements following the stimulation showed an increase in the flow of ground water from the surrounding rock layers. A production test demonstrated correspondingly higher flow and production rates, a first indication that the experiment had been successful. This proved that even moderate pressure stimulation could initiate fissures in these rock strata.

After this initial success, we carried out further experiments in the open, uncased section of the deepest segment of the borehole, which were rather risky. One of the risks was the insertion of a so-called packer at a depth of over 4 km (2.5 mi). A packer is a sealing device for the injection pipe. In principle, it functions like a stopper in an opened champagne bottle, which increases its thickness on screwing it together or by pressure from a lever, and thereby forms a leak-tight plug in the neck of the bottle. Not only was the insertion of the packer important; after the exper-

FIG. 4 | FMI FISSURE IMAGE

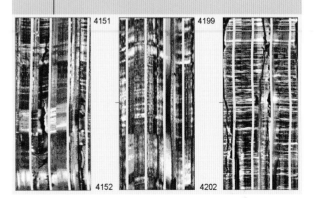

The FMI measurements in November 2003 in the borehole at Groß Schönebeck 3/90 exhibit the "rolled out" wall of the borehole. In them, between 4,150 and 4,200 m (13,600-13,800 ft) depth, one can discern a fissure which was opened by the first massive waterfrac treatment along a length of 120 m (400 ft). The color scale runs from high electrical resistance (light color) to a low resistance (dark).

FIG. 5 | INCREASING PRODUCTIVITY

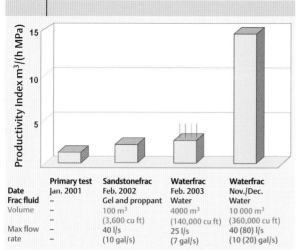

Date	Primary test Jan. 2001	Sandstonefrac Feb. 2002	Waterfrac Feb. 2003	Waterfrac Nov./Dec.
Frac fluid	–	Gel and proppant	Water	Water
Volume	–	100 m³ (3,600 cu ft)	4000 m³ (140,000 cu ft)	10 000 m³ (360,000 cu ft)
Max flow rate	–	40 l/s (10 gal/s)	25 l/s (7 gal/s)	40 (80) l/s (10 (20) gal/s)

The effects of the stimulation. After each stimulation treatment, the productivity was tested and the indices shown here were calculated. The productivity index from February 2003 was obtained as a minimum value; it was probably higher. The value from December was measured at the fissure closing pressure.

iment, during which the system was under considerable strain, it had to be successfully removed again. Had it remained at depth, it would have prevented our further work.

Then, in order to estimate the hydraulic parameters which had been modified by the stimulation, a long-term pumping test over about two months and a long-term pumping test with a production volume of all together 700 m³ (25,000 cu ft) of thermal water were carried out. Over this period of time, we measured the throughput of the different rock strata under moderate intake conditions, the extent of the reservoir, and the chemical composition of the deep-well water which was being pumped up. By comparison with the data from before the first stimulation experiments, we could estimate how the productivity of the sandstone had been affected by the stimulation. The productivity which we obtained after the sandstone stimulation however proved to be still not sufficient for economically competitive electric power generation. We therefore decided to continue the experiments with a massive stimulation.

Massive Stimulation Experiments

In the course of the massive stimulation experiments, all together 14,000 m³ (500,000 cu ft) of water with successively increasing injection rates and increasing pressure were injected below ground. At the highest injection rates, the pressure was far above the earlier value of 17 Mpa (2.3 Gtorr). In order to obtain high flow rates of up to 80 l/s (20 gal/s), we set up a special high-power pumping system. The components of the installation above ground, with the drill head and the connecting lines, had to withstand a pressure of 50 Mpa (6.7 Gtorr), corresponding to 500 times more than atmospheric pressure. The water was pumped out of three

80 meters (260 ft) deep surface wells and was stored in holding basins with a capacity of 1,500 m³ (54,000 cu ft). There, it was chemically treated to make it compatible with the rock and the water at depth – e.g. it was acidified to prevent precipitation of iron hydroxides in the deep reservoir.

In January 2003, we started a first stepwise injection test. We increased the injection rate initially up to 24 l/s (6 gal/s), where the pressure stabilized at 17 Mpa (2.3 Gtorr). We were able to observe that already above an injection rate of 8 l/s (2 gal/s), the pressure increase with further increases in injection rate became smaller. Correspondingly, the injectivity increased: More fluid was pressed below ground per unit pressure; therefore, fissures and crevices were already opening.

We then investigated whether or not the increase in injectivity was also accompanied by an increase in productivity. During a test lasting five hours, we were able to pump 250 m³ (9,000 cu ft) of water out of the deep well; this was a much higher productivity than was found during the pumping tests in the previous Summer 2002. Through physical borehole measurements, which recorded electrical, seismic and radioactive properties, we carried out investigations of the structural rock strata properties in the open segment of the well. A special observation method involved the use of a formation microimager (FMI), which can determine the electrical resistance of the wall of the borehole with a spatial resolution in the centimeter range. It yielded an image which clearly showed a vertical fissure about 150 m (500 ft) long in the lower, uncased segment of the borehole (Figure 4). Further measurements demonstrated that the borehole had passed through the Rotliegend (layer from the earliest Permian era) at its deepest end (Figure 2).

In order to eliminate the risk that the still uncased segment of the borehole would collapse during further massive stimulation experiments in the depth range between 3,985 m and 4,300 m (13,000-14,000 ft), we then installed

Fig. 6 A production test at Groß Schönebeck. The water holding basins can be seen in the photo; they were rapidly filled up during the test.

a protective casing there. This has holes in the so-called storage segment, so that the water can flow through the casing. We then continued the test program in the now secure borehole with stimulation, production tests and stepwise injection (Figure 5).

These tests were able to demonstrate a substantial productivity of about 14 m³/(h MPa) (134 ft³/(h Gtorr)) of water (Figure 6). It follows by calculation from this value that in long-term operation with an input operating pressure of 5 MPa (0.7 Gtorr), at least 70 m³ (2,500 cu ft)of water per hour can be pumped from the deep well. With the current state of technology, the 5 MPa could be produced by an underwater pump, which would have to be installed in the borehole about 500 m (1,600 ft) below the water level. For a normal production, this is a thoroughly realistic scenario. Our experiment was thus able to demonstrate that geothermal power generation in the North German Basin is not only possible, but is also interesting from an energy-economic stand point of view.

Outlook

Fissure stimulation is a first step. Now, it must be proved that the fissure system remains open over a long time period and that it guarantees the transport of a sufficient amount of water. The next step towards geothermal energy production is the successful circulation of water between two spatially separated wells which would be located about 500 meters (1,600 ft) apart in the region of the reservoir. Currently, at Groß Schönebeck a second borehole is being drilled. When it is finished, a circulation experiment lasting several months will be carried out to show whether the fissure system we produced is suitable for the long-term transport and heat exchange of the water below ground: Only production rates which are secured over long periods of time will allow the sustainable exploitation of a hot-water reservoirs; and only then can the investment in power generation be worthwhile.

In Groß Schönebeck, the old deep well is to be used not as a production well but rather as injection well. The experiments up to now have shown that it is ideally suited to this purpose, owing to its injectivity. Furthermore, geometric arguments favor the use of the new borehole being drilled as a production well: it is designed in such a way that it no longer goes down vertically through the storage region, but instead at an angle. This increases its length in this segment, which is decisive for production, and it thus accesses a larger input-flow area.

If a sufficient productivity can be demonstrated, then in cooperation with industrial partners, a research installation for power generation will be installed at Groß Schönebeck. Its purpose is in particular to investigate questions of process engineering, with the economic feasibility of geothermal power generation in the foreground.

The long-term future of geothermal energy in Central Europe can be regarded with optimism. The coupled heat-power geothermal plant which was recently commissioned in Neustadt-Glewe demonstrates that power generation from geothermal heat can be successfully realised under the local geological conditions. The development of geothermal energy in Germany can make an important contribution to the worldwide establishment of renewable energy sources, since the geological underground strata here are typical of Central Europe and thus representative of many regions. Therefore, if this technology operates successfully in Germany, then it can be transferred worldwide to regions of similar geological structure.

Summary

The Earth contains enough heat to permit geothermal electric power generation, which, however requires water temperatures of over 150 °C (300 F). Layers of rocks at this temperature lie at depths of at least 4 km in Central Europe. A power plant must therefore circulate water through deep wells and heat it in the depths of the Earth. In their natural state, however, the rock layers are not usually porous enough to permit this. It is thus planned to force fluids through them at high pressure, artificially enlarging the natural cracks in the rock. This stimulation technology was successfully tested by the GeoForschungsZentrum Potsdam at a 4,309 m (14,137 ft) deep borehole at the German Geothermal Laboratory in Groß Schönebeck. A second borehole is planned to investigate the long-term water circulation at depth. If it is found to be stable, then a demonstration plant will be set up to generate power.

Acknowledgments

For their support of important subprojects, we thank the Federal Ministry for Commerce and Technology (now the BMWA) and the Federal Ministry for the Environment, Natural Conservation and Nuclear Safety.

References

[1] G. W. Hutterer, Status of world geothermal Power Generation, Proc. World Geothermal Congress 2000; Kyushu-Tohoku (Japan) **2000**.

[2] Geothermal Energy Association, Worldwide Contribution of Geothermal Power Generation, www.geo-energy.org

[3] The European Wind Energy Association, Press release on March 10, 2004, www.ewea.org.

About the Author

Dr. Ernst Huenges, physicist and process engineer, is leader of the Geothermal Technology Section at the GeoResearch Center (GeoForschungsZentrum) in Potsdam. Currently, he is chairman of the German Helmholtz Centers for Geothermal Technology. He has participated in many deep drilling projects, for example in the German Continental Scientific Deep – Drilling Program(KTB).

Contact:
Dr. Ernst Huenges,
GeoForschungsZentrum Potsdam,
Telegrafenberg, D-14473 Potsdam, Germany.
huenges@gfz-potsdam.de

The Karlsruhe Process bioliq®

Synthetic Fuels from the Biomass

BY Nicolaus Dahmen | Eckhard Dinjus | Edmund Henrich

*Biofuels could replace a part of the currently-used fossil energy carriers in the near term. To make this possible, raw materials produced over wide- spread areas would have to be made accessible to industrial users of fuels and chemical raw materials on a large scale. The two-stage gasification concept **bioliq** offers a solution to this problem.*

Biomass to Liquid Karlsruhe

Fossil energy carriers form the basis of today's energy supplies. Even though predictions of the time remaining until they are completely exhausted differ widely, there can be no doubt that they will in the long run be used up. As current developments of the prices for petroleum and natural gas on the world market demonstrate, even minor perturbations on a global scale can occasionally produce serious price rises with corresponding negative effects on the world's economy. A consistent utilization of renewable energy sources would alleviate these uncertainties and would at the same time contribute to a reduction of CO_2 emissions into the atmosphere.

While hydroelectric power, geothermal heat, solar energy and wind power are suitable primarily for the production of electric power and space heating, the biomass, uniquely among renewable carbon sources, can play an important role in the production of motor and heating fuels as well as of organic starting materials for chemical synthesis. The biomass of all kinds must therefore be used efficiently.

Biogenic fuels can – even in the short term – replace a portion of the fossil energy sources and thereby make a contribution to the reduction of CO_2 emissions. The aspects mentioned above, together with existing legal and economic requirements, are contributing to increased political and economic pressure to search for solutions. For example, the European Union (EU) requires in its Biogenic Fuel Guidelines that the present proportion of biogenic motor fuels be increased from 2 % of the total consumption in the year 2005 to 5.75 % by the year 2010. The goal for the year 2005 was met, and even surpassed, using the biofuels of the first generation: Biodiesel, plant oils and bioethanol. Their raw materials, vegetable oils, sugar or starch, are produced   from rape, wheat or sugar beets.

A still greater potential for reduction of CO_2 emissions is shown by the fully synthetic biogenic fuels of the second generation, also known as BTL fuels (Biomass-To-Liquids). They can be produced using a broad palette of possible raw materials and employing whole plants. These can be agricultural and forest residues such as straw, waste forest wood, or all the other dry biomass, including energy-yield plants. BTL fuels have the advantage that they are purer and more environmentally friendly than petroleum-based fuels. Furthermore, they can be adjusted to meet special requirements, for example from the automobile manufacturers, and the ever stricter exhaust emission norms. With the distribution infrastructure which is in place today, they can be directly utilized, requiring no new engine technology, and they permit the same vehicle operating ranges as petroleum-based motor fuels.

Obstacles to the Use of the Biomass

In comparison to the use of fossil energy carriers, the production of synthesis gas from the biomass is more complex

INTERNET

Forschungszentrum (Research Centre) Karlsruhe
www.fzk.de/bioliq

European Biofuel Platform
www.biofuelstp.eu

Agency for Renewable Raw Material
www.btl-plattform.de

61

Renewable Energy. Edited by R. Wengenmayr, Th. Bührke. Copyright © 2008 WILEY-VCH Verlag GmbH & Co. KGaA, Weinheim. ISBN 978-3-527-40804-7

FIG. 1 | CONCEPT

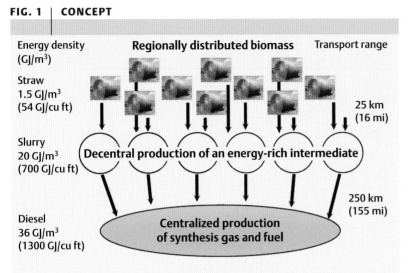

The decentral-centrally deployable bio-slurry gasification concept envisages the production of an energy-rich intermediate, which can be economically transported over longer distances, and which then is converted to synthesis gas and fuels in large, centralized installations.

and more expensive. For the technical application of biomass fuels on a large scale, several hurdles exist.

For one thing, the biomass accumulates over large areas and therefore has to be collected and transported often over long distances. In particular, less valuable biomass such as straw or forest wood residues have a low volumetric energy density (baled straw ca. 2 GJ/m³ (70 GJ/cu ft), in comparison to 36 GJ/m³ (1300 GJ/cu ft) for diesel fuel). Here, the question arises as to the distances over which it is economically and energetically feasible to transport these materials.

FIG. 2 | PROCESS STEPS

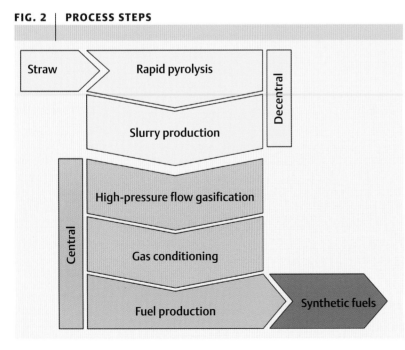

The steps of the bioliq®-process.

In addition, there is a large variety of potentially usable biomass materials. The processes used must guarantee the utilization of the largest possible bandwidth of raw materials. Biomass furthermore consists of heterogeneous solid fuels with to some extent differing chemical compositions; and solid fuels in principle require a greater processing effort.

Making use of already established technologies, e.g. for the processing of fossil-fuel raw materials, helps to shorten the development phase and reduce risks. In particular, the high ash content of many biomass materials causes problems for thermochemical processes, for example due to corrosion or agglutination and blockages of the apparatus. 'Ash content' refers here to the proportion of salts and minerals present.

Motor-fuel synthesis requires a tar-free, low-methane synthesis gas at high pressures of 30 to 80 bar (0.04-0.11 torr), and the extensive elimination of trace impurities which would act as catalyst poisons. On the other hand, this facilitates or even simply allows meeting the stricter exhaust emission norms when the fuel is burned.

Biomass materials, with their average chemical composition of $C_6H_9O_4$, yield on gasification a C/H ratio near to one, which is insufficient for the production of hydrocarbons. This requires an additional process step, the water-gas shift reaction, in which through addition of water, a part of the CO is converted into hydrogen and CO_2. This leads however to poor carbon efficiency. In the long term, it will be expedient to fill the additional hydrogen requirements by utilizing other renewable energy sources.

The biomass is taken from the biosphere, and this must in the long term be done in an ecologically compatible manner. Furthermore, the use of the biomass also has socio-economic aspects, since the new role for arable lands, grasslands and forests as providers or processors of energy-yield raw materials requires the establishment of new logistics, income and labor structures and must not lead to an irreparable loss of food-producing areas.

The Karlsruhe bioliq® Process

At the Karlsruhe Research Center (Karlsruhe, Germany), we have developed a biomass-to-liquid process intended to overcome these logistical and technical hurdles. The emphasis of our process lies in the use of relatively inexpensive, thus far mostly wasted biomass residues. The Karlsruhe synthetic motor fuel is produced by a process involving several steps, the bioliq® process (Figures 1 and 2).

The process steps in the bioliq® Process

1. Rapid pyrolysis: In a first step, the decentrally accumulated biomass is converted through rapid pyrolysis into pyrolysis oil and pyrolysis char (see the infobox "Pyrolysis", p. 64). The air-dried biomass is chopped up and mixed with hot sand, which serves as heating agent, at ambient pressure and under exclusion of air in a double-screw mixing reactor for rapid pyrolysis (Figure 3). The heating, the

FIG. 3 | RAPID PYROLYSIS

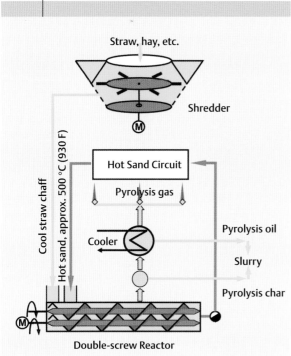

Schematic of the rapid pyrolysis process, which uses a double-screw mixing reactor to produce pyrolysis oil and pyrolysis char, the precursors of bio-slurry.

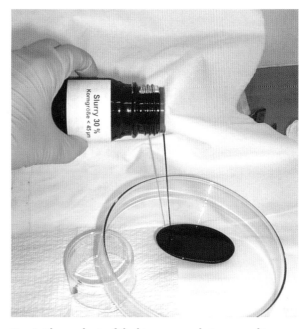

Fig. 4 *The products of the biomass pyrolysis are used to prepare an energy-rich, free-flowing intermediate: bio-slurry.*

actual pyrolytic conversion of the biomass particles at about 500 °C (930 F), and the condensation of the resulting pyrolysis vapors all take place within seconds. Depending on the operating parameters of the reactor and the biomass employed, 40-70 % of an organic condensate (pyrolysis oil) and 15-40 % pyrolysis char are obtained. The remaining product is a non-condensable pyrolysis gas, whose heat of combustion can be used for heating the sand or for drying and preheating of the input materials. The mixing reactor used here was developed about 40 years ago by the industry as a 'sand cracker' for the rapid pyrolysis of various refinery products [1].

2. Slurry production: The brittle and highly porous pyrolysis char is mixed with the pyrolysis oil to give a suspension, called bio-slurry (Figure 4). For this process, the size distribution of the char particles is important. Only when their size is sufficiently small is the resulting slurry stable over long periods of time and can be converted rapidly by the following gasification step. The energy density of the slurry relative to its volume is more than an order of magnitude higher than that of dry straw, and this is an advantage for its transport. Rapid pyrolysis is required at this point in order to obtain the ideal mixing ratio of pyrolysis condensates to pyrolysis char for the preparation of the bio-slurry. This is in turn necessary for a complete utilization of both components.

3. Entrained flow gasification: The bio-slurry is atomised with hot oxygen in a pressurized entrained flow

gasifier and is converted at over 1,200 °C (2,200 F) to a tar-free and methane-poor crude synthesis gas. This flow gasification apparatus originally has been developed for gasification of Central German salty lignite. It is especially well suited for an ash-rich biomass [2]. This is due to a cooling screen onto which the ash precipitates as molten slag and then drains out of the reactor (Figure 5).

The suitability of this type of gasifier was demonstrated in – so far – four test series using different bio-slurries and operating parameters with the 3-5 MW pilot gasifier plant at the Future Energy company in Freiberg. Bio-slurries with up to 33 wt.-% of char were tested, from which a practically tar-free, methane-poor (< 0.1 vol.-%) synthesis gas was obtained. It consists of 43-50 vol.-% carbon monoxide, 20-30 vol.-% hydrogen, and 15-18 vol.-% CO_2. Gasification takes place under a pressure which is dependent on the synthesis to follow. This avoids costly compression of the synthesis gas. Thus, Fischer-Tropsch syntheses require pressures up to 30 bar (0.04 torr), while the methanol or dimethyl-ether synthesis requires up to 80 bar (0.11 torr).

4. Gas purification and conditioning: Before use in a chemical synthesis, the crude synthesis gas must be purified from particles, alkali salts, H_2S, COS, CS_2, HCl, NH_3, and HCN, according to the requirements of the synthesis to follow. This prevents poisoning of the catalysts used for the following synthesis step.

5. Synthesis: The conversion of synthesis gas into motor fuels on a large scale is an established technology. For example, the Sasol company uses the Fischer-Tropsch synthesis to produce more than six million tons of fuel from anthracite coal annually. In this manner, about seven tons of air-dried straw can be used to prepare a ton of synthet-

Pyrolysis (from the Greek: *pyr*, fire; and *lysis*, dissolution) refers to the thermal decomposition of chemical compounds under oxygen exclusion. In this process, depending on the temperature and the processing time, char, liquid condensates (pyrolysis oils) and flammable gases are formed. In rapid pyrolysis, a high yield of liquid products is obtained. This occurs during very short reaction times of a few seconds. This rapid heating for short times is achieved by use of a heat transfer agent such as hot sand, which is mixed intensively in special reactors with the pyrolysis reactants.

ic fuel. Nearly 50 % of the energy originally contained in the biomass remains in the liquid end product. As by-products, heat and electric power can be produced, and they completely meet the energy requirements of the overall process.

Methanol production, with the order of many millions of tons per year, is likewise an established process technology. Methanol is on the one hand an intermediate for a methanol-to-gasoline process. It is however also directly usable itself as a motor fuel. It is used for the synthesis of the anti-knock compound MTBE (methyl tertiary-butyl ether), for the production of rape methylester, and of biodiesel through esterification of rape oil, as well as being an input fuel for high-temperature fuel cells.

FIG. 5 | THE FLOW GASIFIER

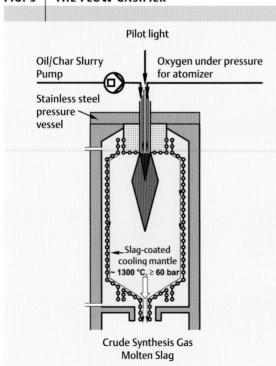

Pilot light

Oil/Char Slurry Pump

Oxygen under pressure for atomizer

Stainless steel pressure vessel

Slag-coated cooling mantle
~ 1300 °C, ≥ 60 bar

Crude Synthesis Gas
Molten Slag

Schematic drawing of the high-pressure entrained flow gasifier, in which the bio-slurry is converted to synthesis gas using pure oxygen at temperatures above 1200 °C.

The Current State of Development

Our work up to now in Karlsruhe demonstrates that even bio-slurries with a high char content resulting from biomass pyrolysis products using pure oxygen in a flow gasifier at high pressures can be completely and safely converted into a tar-free synthesis gas. This process is suitable for practically all materials which yield a sufficiently stable condensate for suspending the char powder after rapid pyrolysis.

Now that the technical feasibility of the process has in principle been demonstrated by experiments with our own and with industrial equipment, the overall process is being further developed as rapidly as possible. For this purpose, we are currently setting up a pilot plant at the Karlsruhe Research Center within the framework of a public grant and with industrial cooperation partners. It will have a biomass throughput of 500 kg/h (approx. 1,100 lb), and is intended to demonstrate and further develop the process, to show the practicability of the procedures applied, to prepare for scaling-up to a commercially relevant size, and to allow the compilation of reliable cost estimates. The first of three constructional phases, the phase of biomass milling, rapid pyrolysis and continuous mixing of the bio-slurry, was authorized and begun in 2005. The pyrolysis plant is operated together with the German firm Lurgi AG, Frankfurt (Main), since 2007. As the next step, the construction of the gasifier and the fuel synthesis plant will follow also in cooperation with Lurgi.

Costs and Development Potential

The Karlsruhe Biomass-to-Liquid process is particularly suited to the requirements of the widely distributed biomass production from agriculture: The rapid pyrolysis and production of the bio-slurry is carried out at a large number of decentrally located plants. They provide the decisive enhancement in energy density needed for further economical transport of the raw materials. The gasification and the following steps of gas conditioning and synthesis can then be performed at a large central installation of a size which makes it commercially cost-effective, and which is supplied by road or rail transport with the bio-slurry raw material.

In a possible scenario, about 40 rapid-pyrolysis installations, each with a capacity of 200 thousand tons of bio-slurry annually, could be set up to supply a central gasification and fuel production plant with a capacity of a million tons of fuel. Then, at a price of 70 Euro (approx. 100 $) per ton for the air-dried starting material, a production price of less than one Euro per kg (less than one $/lb) of fuel could be realized. With integration of the gasifier into an equipment network of the chemical industry, the diversification of the usable products can also be broadened. Along with the option of the utilization of biomass as a source of carbon, necessary in the long term, economically favorable processes could also be developed in the near future.

The focus for the process development is currently on low-grade biomass, which thus far has not been used at all, such as surplus grain straw, barn straw or waste wood. The use of solid wood is not seen as a fruitful solution in the long term. Even though this less problematic starting material might permit the technical realization of the process to be attained more rapidly, it can be expected that the demand for solid wood will increase due to its uses in construction, for cellulose production, and for decentral and household heat and energy production.

The use of entire plants appears to be still more expedient, for example of grain plants or specifically cultivated energy-yield plants. The accompanying systems analysis research [3] leads us to expect an annual production of about 5 million tons of synthetic motor fuel just from the use of waste forest wood and surplus straw, together about 30 million tons of dry material. This corresponds to roughly 10 % of the current consumption of petrol and diesel fuels in Germany.

According to estimates by the Agency for Sustainable Raw Materials (FNR), by the year 2015 a fraction of biogenic motor fuels of 25 % of the overall consumption is possible. Combined with other biochemical and physicochemical processes, a still higher-quality utilization of the biomass in the sense of a biomass refinery should be feasible. Similarly to today's petroleum refineries, it would use a broad spectrum of raw materials to produce a variety of basic chemical materials and fine chemicals, which would result in a clear-cut reduction of the consumption of fossil materials by the chemical industry.

Acknowledgments

We thank the Ministry for Nutrition and Agriculture in Baden-Württemberg (MLR), the German Federal Ministry for Nutrition, Agriculture and Consumer Protection (BMELV), the Agency for Sustainable Raw Materials (FNR), and the European Commission via support of the EU project Renew for their continued support and financing.

Summary

Synthetic fuels from the biomass can provide an important contribution to a renewable energy economy. The Karlsruhe BTL concept bioliq® aims at bringing decentral production in line with centralized processing on an industrial scale. To this end, thermochemical methods are employed: rapid pyrolysis for the production of a readily transportable, energy-rich intermediate product, and entrained-flow gasification to yield synthesis gas and to process it further into the desired fuels. The bioliq process was distinguished with the BlueSky Award by the UN Organization UNIDO in 2006.

References

[1] R.W. Rammler; Oil & Gas Journal, 1981, Nov.9, 291.
[2] M. Schingnitz et al., Fuel Processing Technology 1987,16, 289.
[3] R. L. Espinoza et al., Applied Catalysis A: General 1999, 186, 13 und 41.

About the Authors

Nicolaus Dahmen studied chemistry at the Ruhr University in Bochum, obtained his doctoral degree in 1992 and moved in the same year to the Karlsruhe Research Center. There, he is now concerned with the thermo-chemical transformation of biomass into hydrogen and synthesis gas. As project leader, he is responsible for the construction of the bioliq® pilot plant in Karlsruhe.

Eckhard Dinjus began his studies of chemistry in 1963 at the Friedrich Schiller University in Jena and completed his doctorate there in 1973. In 1989, he obtained the Habilitation, and thereafter he was leader of the Research group "CO₂ chemistry" in the Max-Planck Society. Since 1996, he has been director of the Institute for Technical Chemistry at the Karlsruhe Research Center, and he occupies the chair of the same name at the University of Heidelberg.

Edmund Henrich studied chemistry at the Universities of Mainz and Heidelberg. He received his doctorate in Heidelberg in 1971 and the Habilitation in 1993 in the field of radiochemistry. He has worked at the Karlsruhe Research Center since 1974, and as Division Leader of the Institute for Technical Chemistry there, he is responsible for R and D activities relating to the Karlsruhe BTL process. Since 2005, he has been extraordinary professor at the University of Heidelberg.

Contact:
*Dr. Nicolaus Dahmen, Dr. Eckhardt Dinjus,
Dr. Edmund Henrich, Forschungszentrum Karlsruhe,
Institut für Technische Chemie,
PO box 3640, D-76021 Karlsruhe, Germany.
Nicolaus.Dahmen@itc-cpv.fzk.de*

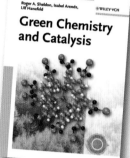

Biogas and Agro-biofuel

Does the Future Belong to Biogas?

BY ROLAND WENGENMAYR

The production of biogas in many small agricultural enterprises provides the great potential of producing heat, electrical power, and even automotive fuel on a predominantly climate-neutral basis. This is particularly attractive in places where farmers can feed the biogas into an existing natural gas grid. In contrast, the climate balance of today's established bioethanol and biodiesel production, relying on the cultivation of high-energy plants, is catastrophic, scientific studies say.

Every agricultural enterprise produces organic waste, e.g., liquid manure from livestock, that can be converted into biogas. Producing methane from this 'resource' is economically interesting and ecologically sensible. Farmers can produce and sell it themselves using small, decentralized biogas plants. In their tanks, bacteria ferment the biogenic material thus releasing methane. The principle is analogous to that in a cow's stomach. Several different types of geologically ancient bacteria cooperate here, all relying on an anaerobic, i.e., oxygen-free environment. While decomposing the biogenic material, they produce a gas containing approximately two thirds of combustible methane and approximately one third of carbon dioxide and some residual gas species. The remaining residue is a high-quality fertilizer replenishing a large number of different nutrients to the soil, previously taken up by the processed plants.

The carbon dioxide released during biogas production is that of a closed CO_2 cycle. The organic waste arising on farms as well as the liquid manure include only such plants that previously withdrew the CO_2 from the atmosphere during growth. When this biogas replaces traditional natural gas or even oil for heating and hot water, it is therefore capable of saving up to 400 grams (approx. 1 lb) of CO_2 per kilowatt hour in the balance of produced thermal energy, German scientists from the Öko-Institut (ecological institute) in Darmstadt, Germany calculated. Hardly any other technology of regenerative energy production features such a high potential of savings on greenhouse gases.

Biogas systems are of particular interest in areas connected to a natural gas grid. In many places in Europe, for example, this infrastructure is fully developed. Farmers can feed their environmentally-friendly produced methane into this grid. However, standard biogas plants need to be upgraded with an additional system: this system is required to increase the concentration in the biogas mixture until it consists nearly exclusively of methane. The gas grid is not only capable of providing space heating but also automotive fuel.

In contrast, the ecobalance of so-called first generation agro-biofuel, i.e. bioethanol and biodiesel, as produced today is very poor. It requires growing high-energy plants such as canola or corn, consuming large amounts of valuable soil and valuable water, and in particular, considerable fertilization. This intense fertilization alone makes today's agro-biofuel a climate killer, Nobel Chemistry Prize winner Paul Crutzen and colleagues have proven scientifically [1]. Laughing gas (dinitrogen monoxide N_2O) released by the nitrogen fertilizer is a 300 times more potent greenhouse gas than carbon dioxide, and for the amounts needed in fertilization, biodiesel, for example, is 1.7 times as harmful in the climate balance as conventional fuel based on crude oil.

If today's agricultural techniques were used to produce biofuel on a large scale, they could severely damage the ecosystem, a recent OECD (Organisation for Economic Co-operation and Development) survey suggests. Geopolitical reasoning directed at striving for independent energy supplies bring forward the only arguments for agro-biofuel as produced today.

This problem may be solved by so-called second generation agro-fuels. They will show a much better ecobalance as they use the complete plant, not only its energy-rich seeds. Moreover, its production will not necessarily require plants that seriously exhaust the soil, as does e.g. corn. A recent study of scientists of the US-Department of Agriculture shows for example that a native prairie grass can be very suitable for energy production: switch grass can deliver five times more energy stored in bioethanol as its production consumes – and it grows without the need of much fertilizer [2]. This results in a promising ecobalance.

References

[1] P. J. Crutzen et al., Atmos. Chem. Phys. Discuss. **2007**, *7*, 11191.
www.atmos-chem-phys-discuss.net/7/11191/2007/acpd-7-11191-2007.html
[2] M. Liebig, K. Vogel et al., Crop Science **2008**, *1*, in print.

About the Author

Roland Wengenmayr is editor of the German physics journal "Physik in unserer Zeit" and a science journalist.

Contact: *Roland Wengenmayr, Physik in unserer Zeit, Konrad-Glatt-Str. 17, D-65929 Frankfurt am Main, Germany. Roland@roland-wengenmayr.de*

Renewable Energy. Edited by R. Wengenmayr, Th. Bührke. Copyright © 2008 WILEY-VCH Verlag GmbH & Co. KGaA, Weinheim. ISBN 978-3-527-40804-7

The Solar Updraft Tower

Electric Power from Hot Air

BY JÖRG SCHLAICH | RUDOLF BERGERMANN | GERHARD WEINREBE

A solar updraft tower power plant combines the greenhouse effect with the chimney effect, in order to obtain electrical energy directly from solar radiation. The plants must be very large to produce energy in an economically competitive fashion. The principle has been proven by building and operating a 200 m (660 ft) high prototype in Spain.

From earliest times, humans have actively made use of solar energy: Greenhouses aid in the cultivation of food crops, the chimney updraft was employed for ventilation and cooling of buildings, and the windmill for grinding grain and pumping water. The three essential components of a solar updraft tower power plant – a hot air collector, chimney, and wind turbines – have thus long been known and used. In the solar updraft tower power plant, they are simply combined in a new way (Figure 1).

Around 1500, Leonardo da Vinci sketched an apparatus which made use of the rising warm air in a chimney to turn a spit for roasting meat. The modern combination with a generator for producing electric power was first described by Hanns Günther more than seventy years ago. Without initially knowing about Günther's publication, we – the Engineering Consultancy Schlaich Bergermann und Partner in Stuttgart – began planning a solar updraft tower power plant. We had gotten the idea while designing and building large cooling towers for thermal power plants.

After initial trials and experiments in a wind tunnel, in 1981/82 we were able to set up an experimental installation in Spain. It operated successfully over a period of seven years. Since then, we have continued in our efforts to realize a commercial solar updraft tower unfortunately thus far without success.

Operating Principles

The principle of the solar updraft tower is shown in Figure 1 (an animation of the operating principle can be viewed, see infobox "Internet" on this page). Under a transparent roof, which is flat, circular, and open around its circumference, the air is warmed by solar radiation (greenhouse effect): together with the natural ground below, the roof forms a hot air collector. At its center, there is a vertical tower tube with large air inlets at its base. The roof is attached to the tower tube by an airtight seam. Since the hot air has a lower density than cool air, it rises in the tower. The resulting updraft draws more hot air from the collector, and at its perimeter, cooler air flows in. The solar radiation thus produces a continuous updraft in the tower. The energy contained in this air flow is converted to mechanical energy using pressure-staged air turbines which are mounted at the base of the tower, and finally, using a generator coupled to the turbines, it is converted to electrical energy.

The solar updraft power plant is technically very similar to a hydroelectric plant – thus far the most successful type of power plant making use of renewable energy sources: The collector roof corresponds to the water reservoir, the tower tube to the penstocks. Both types of power plants utilize pressure-staged turbines, and both yield low power-generating costs due to their extremely long lifetimes and their low operating expenses. The collector roof and the reservoir require comparable areas for the same power output. The collector roof, however, can be set up in arid desert regions and readily removed after its useful life, while as a rule, the reservoirs of power dams flood living (and often even inhabited) land.

Continuous 24-hour operation can be achieved by using water-filled tubes or bags which are laid out on the ground under the collector roof. The water is heated during the day and releases its heat to the air at night (Figure 2). The tubes need to be filled only once; there is no further water consumption. Thus, solar radiation can produce a continuous updraft in the tower.

In order to mathematically describe the time-dependent production of electrical energy from a chimney power plant, a comprehensive thermodynamic and fluid-dynamical model is needed [1]. A good description of the thermodynamics of the chimney power plant as a thermodynamic cycle is to be found in [2]. In the following, we explain the basic relations in simplified form.

Generally speaking, the output power P of a solar updraft tower can be computed as the product of the solar en-

INTERNET

Schlaich Bergermann und Partner
www.sbp.de/en/fla/mittig.html

Solar Updraft Tower Animation Movie
**www.wiley-vch/publish/dt/
books/ISBN978-3-527-40804-7**
(7 MB file)

FIG. 1 | A SOLAR UPDRAFT TOWER

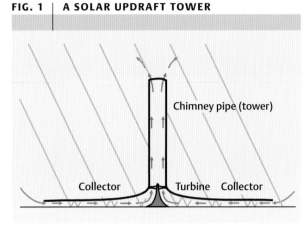

Operating principle of the solar chimney power plant.

FIG. 2 | HEAT STORAGE

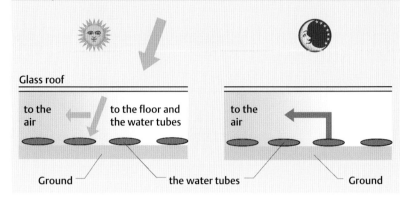

Heat storage using water-filled tubes.

ergy input, $\dot{Q}_{solar}$ multiplied by the various efficiencies η of the collector, the tower, and the turbine(s):

$$P = \dot{Q}_{solar} \cdot \eta_{Plant} = \dot{Q}_{solar} \cdot \eta_{Coll} \cdot \eta_{Tower} \cdot \eta_{Turbine}. \tag{1}$$

The solar energy input into the system can be written as the product of global solar radiation G_h onto a horizontal plane, multiplied by the area of the collector, A_{Coll}:

$$\dot{Q}_{solar} = G_h \cdot A_{Coll}. \tag{2}$$

The tower converts the thermal energy delivered by the collector into mechanical energy. This consists of the kinetic energy of the convection flow and potential energy; the latter corresponds to the pressure drop at the turbine. The density difference between the warm air inside the tower and the cooler ambient air thus acts as the driving force. The column of lighter air in the tower is connected to the surrounding atmosphere at the base of the tower and at its top, and therefore experiences a lift or updraft force. A pressure difference Δp_{tot} is established between the base of the tower and its surroundings:

$$\Delta p_{tot} = g \cdot \int_0^{H_t} (\rho_a - \rho_t) dH. \tag{3}$$

Here, g is the acceleration of gravity, H_t the height of the tower, ρ_a the density of the surrounding air, and ρ_t the density of the air in the tower. Thus, Δp_{tot} increases proportionally to the height of the tower.

The pressure difference Δp_{tot} can be decomposed into a static component Δp_s and a dynamic component Δp_d:

$$\Delta p_{tot} = \Delta p_s + \Delta p_d. \tag{4}$$

Frictional losses have been neglected here. The static pressure difference equals the pressure drop at the turbine; the dynamic component describes the kinetic energy of the flowing air.

Knowing the overall pressure difference Δp_{tot} and the volume flow of the air in the system, i.e. the product of its mean transport velocity in the tower, c_{Tower}, and the tower's cross-sectional area A_{Tower}, we can now compute the power contained in the air flow:

$$P_{tot} = \Delta p_{tot} \cdot c_{Tower} \cdot A_{Tower}. \tag{5}$$

From this, finally, the thermo-mechanical efficiency of the tower can be given as the quotient of the mechanical power in the flow with the thermal current $\dot{Q}_{Tower} = \dot{Q}_{solar} \cdot \eta_{Coll}$, which is input into the system:

$$\eta_{Tower} = \frac{P_{tot}}{\dot{Q}_{Tower}}. \tag{6}$$

The distribution over static and dynamic components in reality depends on how much energy is extracted from the air flow by the turbine. Without a turbine, a maximum flow

FIG. 3 | OUTPUT POWER

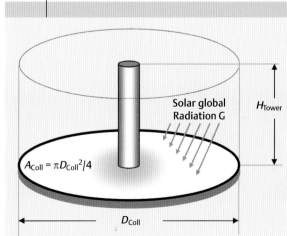

Solar global Radiation G

H_{Tower}

$A_{Coll} = \pi D_{Coll}^2 / 4$

D_{Coll}

The output power of a solar updraft plant is proportional to its collector area and the height of its tower.

velocity $c_{Tower,max}$ is attained, and the overall pressure difference is converted entirely into kinetic energy, i.e. the air flow is accelerated (represents the mass flow rate of the air):

$$P_{ges} = \frac{1}{2} \dot{m} \cdot c_{Tower,\,max}^2. \quad (7)$$

The flow velocity which results from free convection can be determined by making use of the Boussinesq approximation, which summarises the temperature-limited density differences in the air in a "buoyancy term":

$$v_{Tower,\,max} = \sqrt{2 \cdot g \cdot H_{Tower} \cdot \frac{\Delta T}{T_0}}, \quad (8)$$

T_0 represents the ambient temperature at ground level and ΔT the difference between the ambient temperature and that at the inlet of the tower.

With equation (6) and the relation for a stationary state, together with (7) and (8), we obtain the tower efficiency:

$$\eta_{Tower} = \frac{g \cdot H}{c_p \cdot T_0}. \quad (9)$$

This simplified description points up one of the fundamental properties of a solar updraft tower power plant: The efficiency of the tower depends only upon its height.

Equations (2) and (9) show that the electrical power output of the solar updraft tower plant is proportional to its collector area and to the height of the tower: It is thus proportional to the volume of the cylinder which contains both collector and tower (Figure 3). Therefore, a desired power output can be obtained either with a high tower and a smaller collector, or with a large collector and a smaller tower. When frictional losses in the collector are taken into account, the linear dependence between the power and the product (collector area × tower height) is no longer

strictly valid, especially for collectors of large diameter. Nevertheless, it provides a useful rule of thumb.

The Test Installation in Manzanares

After detailed theoretical preliminary investigations and comprehensive wind-tunnel experiments, in 1981/82 we constructed an experimental installation using funds from the German Federal Ministry of Research and Technology, with 50 kW maximum electrical output power. It was located in Manzanares, about 150 km (93 mi) south of Madrid. The test site was put at our disposal by the Spanish energy supplier Union Electrica Fenosa (Figure 4) [3].

This research project was intended to verify the theoretical calculations with measured data and to investigate the influence of individual components on the power output and efficiency of the plant under realistic structural and meteorological conditions. To this end, we constructed a tower of 195 m (640 ft) height and 10 m (33 ft) diameter, surrounded by a collector of 240 m (787 ft) diameter. The plant was equipped with extensive measurement and data collection instruments. More than 180 sensors registered the operating parameters of the system at intervals of a few seconds. The main dimensions and some technical data of the plant are set out in Table 1.

The prototype in Manzanares was planned for an operating period of only three years. For that reason, its chimney was designed as a sheet-metal tube with guy wires, which could be recycled after the end of the experiment. Its wall thickness was only 1.25 mm (0.05 in) (!), and it was stiffened every 4 m (13 ft) by external cantilever trusses. The base of the tower was fixed 10 m (33 ft) above ground level onto a ring. This was supported by eight thin pipes, so that the hot air could flow into the tower with almost no hindrance. In order to provide a streamlined transition between the collector roof and the base of the chimney, a pre-stressed membrane jacket of textile impregnated with plastic was installed (Figure 5).

The tower was supported at four levels and in three directions from the foundations by low-cost thin steel rods. Guy wires, which are usual for such a structure, or even a free-standing concrete tower would have been too costly for

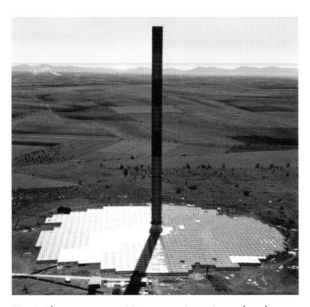

Fig. 4 *The prototype at Manzanares in Spain produced 50 kW of electric power.*

Fig. 5 *The turbine of the prototype plant in Manzanares.*

TAB. 1 | THE PROTOTYPE IN MANZANARES

Tower height	194.6 m (638.5 ft)
Tower radius	5.08 m (16.67 ft)
Mean collector radius	122.0 m (400.3 ft)
Mean roof height	1.85 m (6.07 ft)
Typical temperature rise in the collector, ΔT	20 K (36 F)
Nominal electric power	50 kW
Plastic membrane collector area	40,000 m^2 (430,000 sq ft)
Glass roof collector area	6,000 m^2 (65,000 sq ft)

the limited budget of the project. The sheet-metal tubes were mounted from the ground using an especially developed cyclic lifting procedure. They were lifted in stages using hydraulic presses while at the same time, the bracing rods were adjusted. This was intended to demonstrate that even high towers can be built by only a few skilled workers. Of course, this intentionally temporary construction is not appropriate for a solar updraft tower with a long planned lifetime. For commercial plants, the tower will as a rule be constructed of reinforced concrete.

The collector roof of a solar updraft tower must not only be transparent to sunlight, but also have a long operating lifetime. We therefore tried out different plastic membranes and glass. The experiment was intended to show which material would work best and be most cost-effective on a long-term basis. Glass resisted even violent storms without damage during the operational life of the plant, and proved to be self cleaning; occasional rainfall is sufficient for this. The square plastic membranes were clamped at their edges into profiled channels and were attached at their centers via a plastic plate with a drain opening to the ground. The investment costs for a plastic sheet collector are lower than those for a glass collector roof. However, in Manzanares the sheets became brittle in the course of time and tended to rip. Today, there are more durable plastic materials available, which again make plastic membrane collectors a realistic alternative.

After completion of the construction phase in 1982, the experimental phase began: It was to demonstrate that the principle of the chimney power plant would function under realistic conditions. It was important for us to obtain data on the efficiency of the newly-developed technology. Furthermore, we wanted to demonstrate that the power plant could be operated reliably in a fully automatic mode. Finally, we wanted to record and analyze its operating behavior and the underlying physical processes on a long-term basis.

Figure 6 shows the important operating data for a typical day: Solar radiation, air flow velocity and electrical power output. These clearly show that with this small plant, without a thermal storage system, the electric power output during the day is closely correlated with the solar radiation input (Figure 7).

However, even at night there is some air flow, so that during some of the night-time hours, power can still be gen-

FIG. 6 | OPERATIONAL DATA FROM THE TEST INSTALLATION

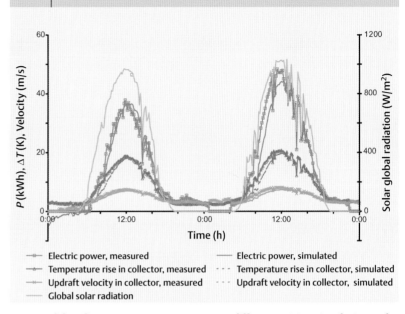

- Electric power, measured
- Temperature rise in collector, measured
- Updraft velocity in collector, measured
- Global solar radiation
- Electric power, simulated
- Temperature rise in collector, simulated
- Updraft velocity in collector, simulated

Measured data from Manzanares: Temperature difference, rising air velocity, and electric power on two days (7th and 8th of June, 1987). Measured output power: 635 kWh; simulated energy output: 626 kWh. No additional heat storage using water-filled tubes.

erated (Figure 6). This effect increases with increasing size of the plant (and thus of the collector), i.e. with increasing thermal inertia of the system. This was confirmed by simulation results for large installations.

During the year 1987, the plant was in operation for 3,197 h, corresponding to an average daily operational period of 8.8 h. As soon as the air flow velocity surpasses a certain value – typically 2.5 m/s (8.2 ft/s) – plant operation

FIG. 7 | POWER YIELD

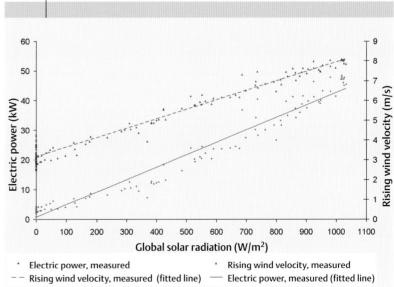

- Electric power, measured
- Rising wind velocity, measured (fitted line)
- Rising wind velocity, measured
- Electric power, measured (fitted line)

The relation between solar radiation and power for the prototype in Manzanares (8th June 1987).

Comparison of the measured (blue) and calculated (by computer simulation, orange) monthly energy production from Manzanares. In the entire year, the plant would yield 44.35 MWh according to the calculation, while the measured value was 44.19 MWh.

FIG. 8 | MONTHLY ENERGY PRODUCTION

Energy yield (kWh/day) — Jan Feb Mar Apr May Jun Jul Aug Sep Oct Nov Dec

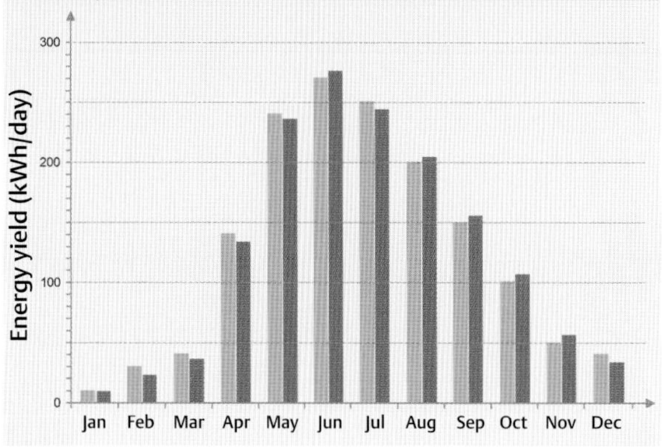

starts automatically and the generator synchronizes with the power grid. The installation as a whole and the individual components operated very reliably.

From the data, we developed a model for a computer simulation. We wanted to obtain a solid understanding of the physical processes and to identify possible starting points for improvements with the aid of the simulations. The computer model describes the individual components, their performance and their dynamic interactions. It is based on a finite-element technique and takes the conservation equations for energy, momentum and mass into account. At present, it is a developmental tool which considers all the known relevant physical effects. Using it, the

thermodynamic behavior of large solar updraft power plants under given weather conditions can be reproduced [4, 5].

Figure 8 gives a comparison between the average monthly energy output as calculated by the simulation and the measured value from the test plant. The two values agree very well.

In summary, we can say that the thermodynamic processes in a solar thermal chimney power plant are well understood. The computational models permit a realistic simulation of the operational behavior of the plant under the given meteorological conditions.

Large Power Plants

Our detailed investigations, supported by extensive wind-tunnel experiments, show the following: The thermodynamic calculations for the collector, the tower and the turbine can be reliably scaled up to a larger magnitude. The small pilot plant in Manzanares covered a much smaller area and contained a much lower volume than for example the 200 MW plant which we will introduce below. Nevertheless, the thermodynamic characteristics of the two plants are astonishingly similar. If we consider for example the temperature increase and the flow velocity in the collector, then in Manzanares, we measured up to 17 K (30.6 F) and 12 m/s (40 ft/s), while the simulation of a 200 MW plant yields average values of 18 K (32.4 F) and 11 m/s (36 ft/s).

Such comparisons support the premise that we can use the measured data from Manzanares and our plant simulation software in order to design large-scale plants. In Figure 9, the results of a simulation for a site in Australia are given. For each season, a period of four days is simulated. This plant is assumed to have an additional heat-storage facility and operates around the clock, in particular also in the Autumn and Winter – then of course with a reduced power output.

Besides other design options, the collector construction with stress ribbons, which proved itself in Manzanares, can also be used in large-scale plants. Ambient wind suction forces can, however, exceed the weight of the light suspended roof structure, so that it upends like an umbrella in the wind (which it in fact did in Manzanares, without damage).

Large plants would require towers of up to 1,000 m (3,300 ft) height. These are a construction challenge, but they are within the realm of possibility today. The CN Tower in Toronto rises to a height of over 550 m (1,800 ft), and the skyscraper Burj Dubai, now under construction, will be more than 700 m (2,300 ft) high, while a skyscraper of more than 800 m (2,600 ft) height is planned in Shanghai. In contrast to a skyscraper, the tower of a solar updraft tower need be only a simple hollow cylinder. It is not particularly slim, and will thus stand securely, while the technical-structural requirements are considerably less strict than those for an inhabited building.

FIG. 9 | A 200 MW UPDRAFT POWER PLANT

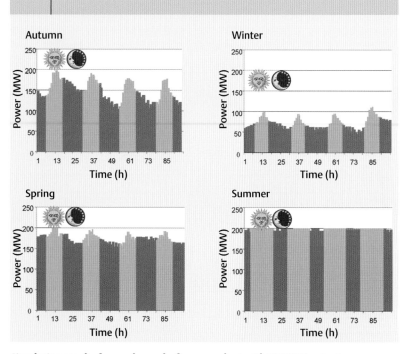

Simulation results for a solar updraft power plant with 200 MW output power, showing the peak-demand output with additional thermal storage capacity. The simulation is for a site far from the equator with strong seasonal variations.

There are various different methods for constructing such a tower: free-standing reinforced concrete tubes, steel sheet tubes supported by guy wires, or cable-net construction with a cladding of sheet metal or membranes. The design procedures for such structures are all well established and have already been utilized for cooling towers; thus, no new developments are required. Detailed static and structural-mechanical investigations have shown that it is expedient to stiffen the tower in several stages, so that a relatively thin wall material will suffice. Our solution is to use bundles of strands in the form of 'flat' spoke wheels which span the cross-sectional area of the tower (Figure 10). This is perhaps the only real novelty in solar updraft towers as compared to existing structures.

For the design of the turbines, we can fall back on experience from hydro and wind power plants, cooling-tower technology, wind-tunnel fans – and of course from the prototype plant in Manzanares. Initially, a single, large vertical-axis turbine in the tower seemed to be the most obvious solution, as suggested in Figure 1 and used in the Manzanares plant (Figure 5). Newer cost estimates have in the meantime convinced us to plan on a large number of horizontal-axis turbines in current designs. These form a ring at the base of the chimney, at the point where the collector is attached to the tower. This allows the use of smaller turbines, which are considerably less expensive. Furthermore, their redundancy guarantees a high availability, since when individual turbines are out of service, the others can continue to generate power. In addition, part-load behavior is improved, since the plantís power output can be controlled by switching on or off individual turbines.

The energy yield of a solar updraft tower is proportional to the global solar radiation, the collector area and the height of the tower. There is thus no physically optimal size, only an economic optimum: The best dimensions for the given site are found in terms of the specific component costs for a suitable collector, tower, and turbines. Thus, plants of different dimensions will be built for different sites – in each case at minimal cost: if collector area is cheap but reinforced concrete is expensive, then one would construct a large collector and a comparatively small tower; and if the collector area is expensive, then one would build a smaller collector and a large tower.

Table 2 gives an overview of the typical dimensions of solar updraft tower power plants. The numbers are based on typical international materials and construction costs. The average cost of labor have been taken here to be 5 €/h (approx. 7 $/h); we are thus considering a site in an emerging nation. The solar updraft tower is especially suitable for such countries. In particular, it becomes clear that the power-generating costs decrease significantly with increasing plant size. In order to achieve an economically viable operation, a solar updraft tower must therefore be of a certain minimum size.

Fig. 10 *Large scale solar updraft tower (artist's impression): the spokes of the internal bracing in the tower, and the ring-shaped visitors' platform with its elevator are clearly seen* (Graphics: SBS = Schlaich Bergermann Solar).

Outlook

The chances for the construction of a large solar updraft tower have never before been so favorable as right now. Nevertheless, it is still not certain whether a solar updraft tower will become a reality in the near future. Here, again, the general dilemma of the first solar updraft power plant holds: Only a very large plant can operate economically. But without the intermediate step of a "small" plant on the megawatt scale, the technical risks to the investors are estimated to be so high that it is difficult to arrange acceptable financing. This challenge must now be met. Once a large solar updraft tower is constructed and on line, additional plants will no doubt follow rapidly. After all, they offer many advantages, since their construction does not involve **consumption** of resources, but simply their **commitment**. Solar updraft tower power plants are constructed essentially from concrete and glass that is from sand and (self-generated) energy. They can therefore reproduce themselves even in the desert – a truly sustainable energy source.

TAB. 2 | TYPICAL BENCHMARK DATA FOR SOLAR UPDRAFT TOWER POWER PLANTS

Name plate power	MW	5	30	100	200
Tower height	m	550	750	1,000	1,000
	(ft)	(1,800)	(2,500)	(3,300)	(3,300)
Tower diameter	m	45	70	110	120
	(ft)	(150)	(230)	(360)	(390)
Collector diameter	m	1,250	2,950	4,300	7,000
	(ft)	(4,100)	(9,680)	(14,100)	(23,000)
Energy output[A]	GWh/a	14	87	320	680
Energy generating cost[B]	€/kWh	0.28	0.16	0.11	0.08
	($/kWh)	(0.4)	(0.23)	(0.16)	(0.12)

[A] At a site with a total global solar radiation of 2,300 kWh/(m2a) (214 kWh/(ft2a)).
[B] With a linear amortization over 20 years and an interest rate of 6%.

Summary

A solar updraft tower power plant combines the greenhouse effect with the chimney effect in order to produce electrical energy from solar radiation. Air is heated beneath a glass roof; it then rises through a central tower and drives turbines while passing upwards. This simple principle can be put into large-scale operation, as was successfully demonstrated by an experimental plant in Manzanares, Spain. However, solar updraft towers must have enormous dimensions in order to generate electricity economically.

References

[1] M. A. Dos Santos Bernardes, A. Vofl, G. Weinrebe, Solar Energy **2004**, 75 (6), 511.

[2] A. J. Gannon und T. W. v. Backström, Thermal and Technical Analyses of Solar Chimneys, in: Proc. of Solar 2000, (Eds.: J. E. Pacheco, M. D. Thornbloom), ASME, New York **2000**.

[3] W. Haaf *et al.*, Solar Energy **1983**, 2, 3.

[4] W. Haaf, Solar Energy **1984**, 2, 141.

[5] G. Weinrebe and W. Schiel, Up-Draught Solar Tower and Down-Draught Energy Tower – A Comparison, in: Proceedings of the ISES Solar World Congress 2001. Adelaide (Australia) **2001**.

About the Authors

Jörg Schlaich, born in 1934, studied architecture and structural engineering at the University of Stuttgart, the TU Berlin and in Cleveland, Ohio. Doctorate 1962. 1963–79 engineer and partner in the firm Leonhardt und Andrä, Stuttgart. 1974-2000 professor and director of the Institute for Solid Structures in Stuttgart. Since 1980, he has been a partner at Schlaich Bergermann und Partner, Stuttgart. Participation in the construction of the Olympic Stadium in Munich. His project team "Think" took second place in the competition for new World Trade Center.

Rudolf Bergermann was born in Düsseldorf in 1941, studied structural engineering in Stuttgart. 1966 – 67 engineer in the firm Ed. Züblin AG, Stuttgart. 1974 – 79 senior engineer, in cooperation with Jörg Schlaich, Leonhardt und Andrä. From 1980 on, partner at Schlaich Bergermann und Partner. Chief designer of many remarkable structures, e.g. Ting Kau bridge in Hong Kong. Honorary doctor of Cottbus University.

Gerhard Weinrebe was born in 1965, studied aeronautics and space technology at the University of Stuttgart. Worked as a researcher at the Plataforma Solar de Almería in Spain. He obtained his doctorate from the University of Stuttgart. Since 2000, he is a member of the Schlaich Bergermann und Partner solar energy team.

Contact:
*Prof. Dr.-Ing. Jörg Schlaich,
Schlaich Bergermann Solar,
Hohenzollernstrafle 1, D-70178 Stuttgart, Germany.
g.weinrebe@sbp.de*

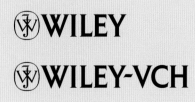

Wave-Motion Power Plants

Energy Reserves from the Oceans

BY KAI-UWE GRAW

An old dream of humanity is to make use of the immeasurable energy of ocean waves. Their destructive power has up to now not allowed any economically reasonable construction to survive for long, although there have been many promising attempts.

The reduction in carbon dioxide emissions which is increasingly pressing and is a major goal of governments and societies, has bestowed an increased importance on wave energy – as on all renewable energy sources. Interest in wave energy power plants, which could make appreciable contributions to the world's energy supply, is steadily growing. The use of wave energy for generating electric power has been under investigation for many decades. However, the countless, sometimes extremely naive suggestions for the application of wave energy have given this renewable energy source a dubious aftertaste in the public perception. But the long-term commitment of a few research teams is now leading to a rethinking of this view.

The ocean waves contain inexhaustible reserves of energy. They are estimated to store around ten million terawatt hours of wave energy per year. This makes them, in principle, very attractive as an energy source. However, large waves can deploy a destructive force which makes huge demands on the stability of the wave power plant under load. It is thus particularly interesting to employ wave power plants precisely where the force of large waves must in any case be broken: along coastal protection installations. Conventional breakwaters only reflect or dissipate the wave energy without making use of it. Wave power plants, in contrast, extract this energy and convert it into useful electric power. Figure 1 shows a wave power plant which protects the harbor of Vizhinjam in India. Wave power plants which convert energy without fulfilling any

Fig. 1 *The Trivandrum OWC is a wave-breaker power plant which was constructed in 1990 at the harbor of Vizhinjam, India. It delivers 150 kW of electrical power (width: 8 m (26 ft), water depth: 10 m (33 ft)). It is a research project under the auspices of the Indian Institute of Technology.*

protective function can even take the form of free-floating installations in the ocean (Figure 2).

The deployment of this technology for coastal protection plays in important role in particular in India and in Japan. In the U.S.A. and Europe, it was at first investigated only very hesitantly. In general, most wave energy is available on the Western coasts of the big oceans, i.e. on the North and South Pacific coasts of America, on the Atlantic coasts of Europe and (South) Africa, on the Indian Ocean coast of Australia, and on the South Western coast of New Zealand. Precisely for this reason, the coupling of energy extraction with coastal protection is one of the few possibilities of making the use of wave energy economically attractive. Large coastal protection facilities however are not without side effects on the ecological system; but there has as yet been no research on this subject.

The Formation and Propagation of "Gravity Waves "

The major portion of the energy which is stored in ocean waves is transported by so-called gravity waves (Figure 3). They are initiated by the wind and their motion is governed almost entirely by gravitation. Figure 3 also illustrates the forces which activate the waves: short to medium waves are mainly produced by the wind, longer ones by air-pressure

wave crests form foamy whitecaps. Turbulences consume part of the wave energy. When the waves have reached their maximum height and their period no longer changes, even when the wind continues to blow, i.e. when a stationary state has been attained, the sea condition is called a "fully developed sea". After the wind has died down, the waves can maintain their energy over distances of many thousand kilometers. They are then referred to as groundswell.

When the waves move into shallow water, their length and velocity decrease. Friction with the ocean bottom dissipates their energy and often changes their direction. When the wave velocity has dropped to a certain limiting value, the waves break and form a foamy surf. The wave breaking also leads to energy loss through turbulence. In planning coastal wave energy power plants, this process must be taken into account.

◄ **Fig. 2** *The MIGHTY-WHALE OWC of Japan's Marine Science and Technology Center (JAMSTEC) was commissioned in 1998. this pilot plant is 30 m (100 ft) wide and floats in water at a depth of 40 m (130 ft); its output power is 110 kW.*

The Basic Technology for Exploiting Wave Energy

Beginning in 1986, a simply-constructed wave power plant was built in Norway, where it was operated for about twelve years. The plant was on the island of Toftestallen near Bergen, and it was intended as a demonstration project for interested groups. The TAPCHAN (TAPered CHANnel) directs the water from the incoming waves into a channel which rises and narrows, then empties into a raised basin. The water then flows steadily from this reservoir back to the ocean. In the process, it can power a conventional low-pressure turbine. The channel of the prototype plant had a 60 m (200 ft) wide opening on the incoming wave side and was between 6 und 7 m (20-23 ft) deep. The reservoir was at a height of 3 m (10 ft) above sea level. The incoming waves were steepened by the trumpet shape of the chan-

differences due to weather fronts or by earthquakes (tsunamis); extremely long waves are due to the tidal forces. The figure also shows the three forces which in general determine the propagation of the waves: surface tension, gravitation, and the Coriolis force. This last force is due to the Earth's rotation; it is weak and has a noticeable influence on waves only when they are several kilometers long. The surface tension of the water is also a very weak force. It is important only for waves that are shorter than about one centimeter (about half an inch): such waves are deformed by the surface tension. In all other cases, gravity waves predominate. The force of gravity pulls the water in the wave crests down towards the troughs and thus tends to equalize the differences in height.

Water waves produced by the wind are generated mainly over deep water. Their shape depends on the wind velocity, the duration of the wind, and the distance they have propagated since they were formed. The regions of the ocean surface with the most wave energy are therefore the open oceans far from the equator (Figure 4). The wind is subject to friction with the surface of the ocean water, pushes on individual water particles and thus accelerates the water layers near the surface. Turbulences in the airflow give rise to pressure differences between different parts of the water surface. To equalize these differences, the surface rises and sinks. This now rough surface is subject to ever stronger pressure differences from the wind, which in turn increase the amplitude of the surface roughness. In this way, higher and higher, quasi periodic waves are formed.

Wave dynamics in the end limit further growth of the waves. The simple model of "linear wave theory" already gives a realistic value for the maximum wave height. It is ca. 14 % of the wavelength. According to linear wave theory, the individual water particles in the wave attain speeds at the tops of the wave crests which are greater than the propagation velocity of the waves. They practically "fall out" of the wave in the direction in which it is moving. At this maximum height, the wave thus becomes unstable, and the

FIG. 3 | WAVES AND THEIR ENERGY

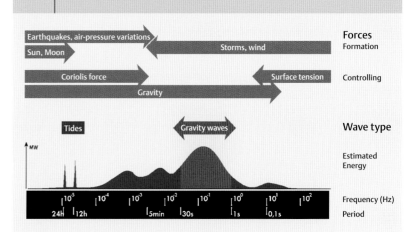

The energy distribution of waves which act at the seacoasts, as a function of their oscillation period; gravity waves (red) contain the major portion of the energy. Their period lies between one and thirty seconds. Above, the forces are indicated which are responsible for the formation and propagation of the various types of waves.

FIG. 4 | DISTRIBUTION OF THE AVERAGE WAVE ENERGY

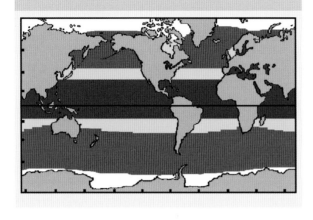

nel, so that they overflowed its banks, allowing the water to flow from the sides of the channel into the reservoir. The reservoir was lower than the end of the channel, preventing the water from flowing back out.

The TAPCHAN in practice even exceeded by a small amount its planned output of 350 kW maximum power and GWh annual energy production; this is so far a notable exception for wave energy power plants. Operating problems arose from earthslides after strong rainfall, pieces of rock flushed into the channel by the sea, and damage to the channel walls. In contrast to these constructional difficulties, the power generation with a standard water turbine caused no problems. However, there were no follow-up land based projects – probably because reserving large coastal regions for such storage power plants would not be economically feasible. Coastal regions are exploited today in so many diverse ways that the integration of new, area-intensive uses would hardly be enforceable.

The principle anyways goes offshore: The Danish Wave Dragon® is a floating, slack-moored energy converter of the overtopping type that uses basically the same physical principle as TAPCHAN. The first prototype connected to the

grid is currently deployed in Nissum Bredning, Denmark. Wave Dragon basically consists of two long wave reflectors focusing the waves towards a ramp. Behind the ramp there is a large reservoir where the water that runs up the ramp is collected and temporarily stored. The water leaves the reservoir through hydro turbines that utilize the head between the level of the reservoir and the sea level.

Today's Standard Technology: the OWC

OWC is an abbreviation for Oscillating Water Column and describes a construction which makes use of the wave motion in rising and falling water columns. Figure 5a shows how a typical OWC functions. It consists of a chamber with two openings. One of these is on the side of the incoming waves and lies beneath the water level. The water can enter the chamber through this opening, driven by the wave energy. The second opening allows the pressure to equalize with the surrounding air. The water column in the chamber moves up and down at the wave frequency, and thus "breathes" air in and out through the second opening. This "breath" drives an air turbine. It is constructed in such a way that it converts the oscillating motion of the air column into continuous rotational motion (see the infobox "Operating Principles of OWC Turbines" on p. 81).

In principle, an OWC represents a simple motion-conversion device (Figure 5b): In order to drive the generator, it converts the strong force at low velocity of the wave motion into a motion of the air column with a weak force but a high velocity. The essential aspect is that the low specific mass of the air permits a high acceleration.

OWC systems have been in use for decades for the energy supply of beacon buoys (Figure 6). They were invented by the Japanese Yoshio Masuda. In an OWC buoy, a vertical pipe assumes the function of the chamber. It reaches down into the calmer water layers below the buoy. Therefore, the water column in the pipe is at rest relative to the waves outside – but it moves relative to the buoy, since the latter is raised and lowered by the wave motion. Like standard OWCs, most buoys employ an air turbine. Such buoys

FIG. 5 | THE OWC PRINCIPLE

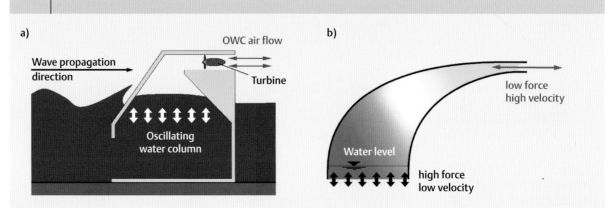

have become accepted rather quickly for applications that require limited power outputs. Some of them have survived more than twenty years of operation at sea.

Existing OWC Projects

Between 1978 and 1986, in an international experiment initiated by Japan, an OWC system was tested for the first time on a large scale: the ship 'Kamai' carried out three series of tests in the Japan Sea in which turbines with up to a megawatt of power output were installed. Only in 1998 was the idea of a floating OWC again taken up. The new Japanese prototype 'Mighty Whale' however has a power output of only 110 kW (Figure 2).

It has taken a similarly long time until the first test of an OWC constructed at a coastal site with a relatively high power output was pursued. From 1985 to 1988, the Kværner company in Toftestallen tested an OWC built on the rocky coastline with a power output of 0.5 MW; it was constructed mainly of steel. Only since the end of the year 2000, on the island of Islay in Scotland, has once again a coastal OWC with a similar power output been in operation: LIMPET (Land Installed Marine Powered Energy Transformer). It is constructed by the Scottisch firm Wavegen for the most part of concrete with 250 kW output power, and is a very similar project on the Azores island Pico. In 2005 Wavegen became a subsidiary of the big German company Voith Siemens Hydro Power Generation that currently builds a small breakwater installation in Mutriku (Spain) and plans a 3.6 Megawatt plant with 36 turbines in Lewis (Scotland). Wavegen announced furthermore that two projects, one in the U.S.A. (Reedsport, Douglas County, Oregon) and on the Canary Islands (Europe) are under development. All those projects are concrete constructions as has been tested already since 1983 in Sanze (Japan), since 1990 on the island of Dawanshan (China), and since 1988 on the island of Islay with considerably smaller prototype plants. Whether concrete will prove to be more enduring than steel is still under debate by the experts.

Breakwater construction usually consists of many concrete cubes of the size of a single or multiple-family house. For breakwater wave energy power plants, one or more such cubes are modified in such a way that they can be employed as OWC chambers. The construction of these OWC chambers directly at the locations where they are to be used has proved to be the main problem for all of the test projects mentioned. At a location where the waves break onto the shore, people have to work continuously for several months – a dangerous, difficult and therefore expensive undertaking. For this reason, those construction engineers who built OWCs near Trivandrum (India) in 1990 and Sakata (Japan) in 1988 took a different route: Both were built as concrete caissons. These caissons were produced using established methods in a drydock. The firm ART (Applied Research and Technology) also built its steel wave power plant OSREY (Ocean Swell Powered Renewable EnergY) with 500 kW of output power at a shipyard. ART-OSREY demon-

strated that even the installation of a previously-constructed wave breaker has its uncertainties: during the installation of the power plant in water 20 m (66 ft) deep off the Scottish coast, a severe storm came up. The construction was not designed to withstand such stresses during its installation phase, and it was destroyed.

All of the OWCs so far existing have to be classed as test installations, with which construction techniques can be tried out and turbine technologies developed – although Limpet and the Indian OWC have already fed power into the local grid. ART-OSREY was supposed to be the first commercially-operating OWC prototype. This role has been taken over by LIMPET (Figure 7).

Technological and Economic Questions

Why does wave power not come "from the socket" yet? As we have already described, the installation of the power plants in the rough locations where they are to be operated is difficult. Their construction must guarantee a long operating life. A further hurdle is the developement of turbines which are suitable for OWCs. The turbines used up to now do not operate well: Their efficiencies are too low and their constant velocity operation is unsatisfactory. Wells turbines so far achieve efficiencies only in the range of 50 bis 70 %. Conventional turbines, in contrast, operate with up to 90 % efficiency. Even though they produce electrical power in only one flow direction, they can still be considered serious competition for new designs. To answer this question, we are carrying out a research project a the University of Leipzig, supported by the German Federal Environment Foundation.

OWC designs have also been tested which supply conventional turbines with a uniform air flow and thereby com-

FIG. 6 | AN OWC BUOY

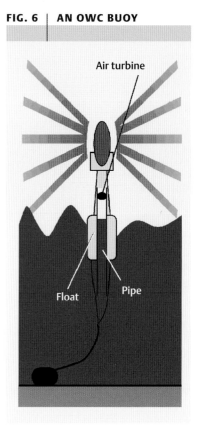

OWC buoys are for the most part designed as beacon buoys, with their own autonomous energy supplies.

Fig. 7 *LIMPET is a coastal OWC with an output power of 250 kW, that has been in operation on the island of Islay (Scotland) since the year 2000 (Photo: Wavegen).*

Fig. 8 *The Archimedes Waveswing consists of submerged cylinders which "float" in the water. The outer cylinder (green) moves up and down relative to the central one (black). To generate power, linear generators (grey) utilize this motion – reversing the principle from that of a magnetic levitation railway.*

Fig. 9 *The prototype of a Pelamis plant, which was installed in 2004 off the Scottish Orkney Islands* (Photo: Pelamis Wave Power Ltd.).

pensate their disadvantages. For example, the 30-kW Kujukuri OWC, which was built in 1987 in Japan in the Kujukuri harbor, uses pressure storage vessels for the air which is compressed by the waves. The storage vessels supply conventional turbogenerators without reversing the air flow.

An additional technical problem is the quality of the electrical "wave power": It fluctuates, and the fluctuations must be compensated by the power grid. As in the case of wind power plants, the power production varies with changing weather conditions, and depending on the location of the plant, the tidal variations add to the fluctuations. For OWC power plants, there is in addition a periodic fluctuation which reflects the relatively high frequency of the incoming waves and is passed on to the power grid.

To be economically feasible, power plants must be planned for an operating lifetime of at least twenty years, while their "moving parts" should last at least ten years. In estimating the financial boundary conditions for the use of wave energy, a fundamental physical property of the waves must be considered: The waves' energy increases as the square of their amplitude or height. To illustrate the economic and technical restrictions,

let us consider an example. Suppose that a wave energy converter is designed to extract energy from waves that are one meter (3.3 ft) high. In order to withstand extreme storms, however, at the same time it has to be able to deal with waves that are roughly ten times higher – that is, waves ten meters high. Such waves carry wave energies which are a hundred times greater than that of the waves for which the plant is designed! This requirement can cause the construction costs to explode in comparison to those of other types of power plants.

Other Technologies for the Exploitation of Wave Energy

As mentioned at the beginning, there are numerous different ideas for extracting energy from ocean waves. Two of them are distinguished by their special construction. Their developers are trying to take the bull by the horns and solve the main economic problem of wave energy plants: the requirement that the plant be able to survive "monster waves". Both systems are currently being constructed as prototypes.

Figure 8 shows the Archimedes Waveswing. The pilot plant with a peak power of around 1.5 MW was installed off the coast of Portugal in 2004 and already tested with grid connection. It consists of two submerged cylinders which are anchored to the ocean floor. They are 21 m high (69 ft) and 10 m (33 ft) in diameter. A wave which passes above this construction gives rise to a "dynamic" variation in the buoyancy; the oscillating water flow moves the closed cylinders up and down. This construction has the advantage that it is not subject to the strong wave forces near the surface of the water. Its disadvantage: in the deeper water levels, only a small fraction of the wave energy is still available. If the cylinders were anchored near the surface in shallow water, then the rolling motion of the waves would cause strong horizontal motions; the cylinders cannot convert these mo-

tions into usable energy, they only add to the load on the structure. Archimedes Waveswing (AWS) produced so far a continuous average output of up to 1 MW, announced its constructor AWS Ocean Energy, and this firm plans to install by 2010 utility-scale AWS machines.

The builders of Pelamis (Figure 9) asked themselves the question of how an energy converter would have to be designed in order to survive large waves with the lowest construction costs. The result is a system which can be strongly deformed by smaller waves, and converts their energy in an optimal fashion, but still will not be damaged by large waves. Since Pelamis is a kind of "snake" made of movable, coupled segments, it was given the name of a genus of sea snakes. With a total output power of 750 kW, the prototype Pelamis is around 120 m (400 ft) long and contains three segments each of which has a diameter of 3.5 m (11.5 ft).

Normal sea conditions, with a relatively moderate vertical oscillation, cause the segments of Pelamis to perform a horizontal evasive motion, like a sea snake swimming. Hydraulic assemblies convert this motion into usable energy. They produce a high pressure whose energy is stored in intermediate air pressure chambers. These chambers then feed hydraulic turbines which operate at constant speed. Pelamis thus converts brief wave impulses with a high power into a constant, lower power from the turbines. For this reason, the pumping power capacity of the hydraulic pumps must be designed to be considerably greater than the power output of the hydraulic turbines.

With respect to large waves, in contrast, Pelamis acts like a stiff structure. It can not follow the large vertical oscillations, and simply dives through the waves. Since its structure has a relatively small cross-sectional area, it is subject to only moderate forces when diving through large waves. For this reason, this new approach is very promising. Immediately following the first successful tests of the prototype at Orkney Islands, a first commercial Pelamis unit is now planned to be installed in a wave-energy farm off Portugal's coast in 2010.

Can Wave Energy Soon Be Commercially Exploited?

The energy crisis in the 1970's aroused strong interest in renewable energy sources. Thus, the use of wave energy for electric power generation, like many other renewable energy sources, was the subject of intense research. After the end of the crisis, the technology was however deemed too expensive and was put aside. Only a few, for the most part Asiatic research institutions continued working on the topic. In Europe, this situation changed only about ten years ago. At that time, the European Union included wave-energy conversion in the research program JOULE. A major part of the European prototypes described here are based upon this research support. These activities are distributed rather inhomogeneously over the various EU countries. In Germany, only recently has there been an official program of research support, currently an OWC demonstrator project

OPERATING PRINCIPLES OF OWC TURBINES

In wave power plants, the turbines have to withstand about twenty load alterations every minute. In OWC power plants, these vary not only between zero and maximum flow rates, but they also periodically reverse the direction of flow. The development of frequency transverters for wind turbines has made it possible that the generators no longer require a constant rotation speed. But so far, the direction of rotation must be kept constant. In order to solve this problem, there are two approaches in use today: the Wells turbine and the impulse turbine. Both types are currently being tested to determine whether they are serviceable for wave power plants.

Wells Turbines

The Wells turbine has blades with a symmetric profile which is perpendicular to the air flow. Once they are set in motion, they maintain their direction of rotation, even if the direction of the air flow changes. When the turbine is rotating, the overall approach flow to the blades is composed of two components: one component is the air which is flowing into or out of the OWC, the other is the approach flow of the airfoil, which depends on its rotational speed. The two components add to give an overall flow. It makes an angle with the airfoil which depends on the two flow velocities. The resulting reaction force on the airfoil is perpendicular to the overall flow. The reaction force can be

further decomposed into a driving force and a buoyancy force.

The buoyancy force simply pushes against the turbine bearings and is not useful; only the driving force contributes to the rotation of the turbine. If the direction of the OWC air flow reverses, the force diagram remains the same. This decomposition of forces makes it clear why a Wells turbine has a lower efficiency than a conventional turbine.

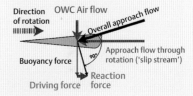

Distribution of forces on one blade of a Wells turbine.

Impulse Turbines

Those designs which redirect the approach flow within the turbine rotor are much simpler. Impulse turbines have for this purpose fixed guide vanes (shown black in the figure), between which the turbine rotor (blue) revolves. The figure shows (as dashed red lines) how the guide vanes direct the OWC air flow in such a way that it produces a driving force on the blades of the turbine rotor. Reversal of the flow direction does not change the direction of rotation of the rotor. The impulse turbine has the disadvantage that losses occur in the redirection of flow before and after the turbine rotor, as well as in the gaps between the turbine blades and the guide vanes. For this reason, it initially could not compete successfully with the Wells turbine. Its superior constant-velocity properties could however soon put it back on the map.

OWC Air flow

Direction of rotation

Operating principle of the impulse turbine.

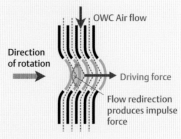

Operating principle of the Wells turbine.

is under review. In Denmark, on the other hand, a broad-based program is in place, which was intended to repeat the successes achieved in the use of wind energy. In the Atlantic countries of Europe, in particular in Portugal, Spain and in the United Kingdom, efforts have been increased strongly as a result of the CO_2 discussion.

This European research is today most certainly the motor for further developments: It started by carefully examining the available technologies, carried out feasibility studies for particular power plants, and prepared a survey of the usable ocean-wave energy on the European seacoasts. In the meantime, several European pilot projects are operating or are almost complete. If these projects can demonstrate that they are technologically operable, then commercial applications will be possible. In Asia, the research projects have thus far not succeeded. The Japanese energy suppliers are not willing to feed the fluctuating electrical power from wave energy into their power grid; in India, the energy production costs are expected not to exceed those of conventional power generation. Both requirements have already been discussed in Europe during the introduction of wind energy plants, and to a large extent they have been resolved. For example, it has been found in practice that the power grid can buffer out the variations from wind energy without problems. The political decision that a comparison with conventional power generation will have to include all the external costs – such as the environmental effects of the waste products – has also been very helpful. This has for the first time created economically acceptable boundary conditions for the commercial wave-energy projects.

How Expensive Would Wave Energy Be?

A rough estimate of the most important cost factors yielded the following results: The price per kilowatt hour would be ca. 10 Euro-cents/kWh (approx. 14 $-cents/kWh), about on the same level as the price for wind energy. This estimate already takes into account a number of uncertainties, among them factors such as the size of the future power plants, the state of development of their systems, and the evolution of financing costs (bank interest rates). According to this estimate, the exploitation of wave energy would be clearly cheaper than solar power, which at present is estimated at 0.5 to 1 Euro/kWh (0.7-1.4 $/kWh). The price is in fact strongly dependent on the costs of underwater power cables, which are required to transport the current generated. The increasing number of offshore wind energy parks will certainly lower the production costs of these cables. Furthermore, combined wave and wind energy power plants could utilize the same underwater cables cooperatively. This would effectively halve the cable costs. An additional argument in favor of the use of ocean wave energy is the European electric power consumption practices: In Winter, power consumption increases. At the same time, the Pacific and Atlantic weather labs produce more wave and wind energy on the North American and European seacoasts. In comparison to solar energy, which is for the most part available in the Summer, these two energy sources thus conform much more closely to the seasonal energy consumption patterns in North America and Europe.

Conclusions

The technological and commercial boundary conditions today – for the first time – are bringing the use of wave energy into reach. For European energy policy, it will be decisive whether the current prototype plants operate reliably in the coming five to ten years. Success here would stimulate the construction of commercial wave power plants. With an increasing number of plants, the technology could be optimized and would certainly become less expensive. Then, the immeasurable energy of the oceans could provide a perceptible proportion of the energy supply for humanity on a long-term basis.

Summary

Prototypes of wave-motion power plants have been tested for several decades, mostly utilizing the technology of an oscillating water column (OWC). Autonomous OWC lighthouse buoys have already become successfully established. Larger power plants have been tested in the form of floating and stationary prototypes with power outputs of up to a megawatt. The main problem and most serious cost factor are extreme waves, which can destroy such systems during installation and operation. The European Union is now supporting pilot projects with the goal of introducing commercial wave-motion plants if the pilot plants are successful. Electrical power from wave-motion plants would presumably cost about the same as wind-generated power.

About the Author

Kai-Uwe Graw, born in 1957 in Berlin, obtained his doctorate in Structural Engineering in 1988, at the Technical University of Berlin. He received his postdoctoral lecturing qualification for civil engineering in 1995 at the Bergische Universität – Gesamthochschule Wuppertal. From 1996, he was professor for substructural and hydraulic engineering at the University of Leipzig, and since 2005, at the Technical University of Dresden.

Contact:
Prof. Dr.-Ing. habil. Kai-Uwe Graw,
Institut für Wasserbau und Technische Hydromechanik, Technische Universität Dresden,
D-01062 Dresden, Germany.
kai-uwe-graw@tu-dresden.de

Hydrogen as a Carrier of Energy

Is the Hydrogen Economy Around the Corner?

BY GERD EISENBEISS

Hydrogen is often touted in the media and by politicians as a miracle solution, offering an environmentally-friendly and inexhaustible energy supply for the future. What can we realistically expect from this energy carrier?

Hydrogen is once again much under discussion. This has often been the case during the history of energy technology and its visions. Today, however, hydrogen is part of the vocabulary of statesmen, and not just a topic for visionary thinkers, who for example see the Sahara as an inexhaustible source of hydrogen energy. There, according to the ideas of physicist Reinhard Dahlberg some 50 years ago, a solar breeder process could be set in motion. The solar radiation would first be utilized to produce pure silicon from the sand. This silicon would then, in the form of photovoltaic power plants, deliver power not only for continued silicon production, but also for hydrogen generation by electrolysis apparatus. The hydrogen could then be exported to Europe. Today, US president George W. Bush – as well as the past EU Commission president Romano Prodi - see hydrogen as the key to the solution of future energy-supply problems. Concepts such as the **hydrogen economy** or even the **hydrogen society** are employed to emphasize the central importance of this energy carrier.

This chapter is intended to give a serious discussion of which problems can in fact be solved by using hydrogen. As we shall see in the end, it is indeed highly probable that hydrogen will be utilized for applications in transportation as one element of the energy economy. For this to occur, it is however decisive that hydrogen be available at lower cost and in larger amounts than the hydrocarbons petroleum and natural gas, which will become scarcer in the future.

That time is still relatively distant, so that one must warn against premature construction of expensive infrastructures for a future hydrogen economy. On the other hand, it is reasonable to carry out thorough research into the open questions regarding hydrogen technology and to demonstrate possible future technical solutions.

Hydrogen – the Solution to Which Problems?

At a time when the increasing cost of petroleum is being painfully perceived by the public, it must be reminded again and again that precisely this problem cannot be solved by hydrogen. Hydrogen cannot replace petroleum, which is becoming scarcer (Table 1). The reason for this is that hydrogen is only a secondary energy carrier, like electric power; that is, it must be generated, like electricity, by using a primary energy source. As a result, it cannot replace the limited oil and gas reserves; this can be done only by coal, nuclear energy or renewable energies. Hydrogen must be generated by using one of these energy sources.

The world's largest hydrogen service station was opened in Berlin in 2004. It is a demonstration project of the Clean Energy Partnership (www.cep-berlin.de). (Photo: CEP.)

The future energy provisioning is however threatened not only by a crisis of scarcity on the supply side, but also by a crisis in the disposal of the waste products from energy consumption. For example, nuclear energy is not used, or its use is being terminated, in a number of countries because – not least – of the fact that the disposal problems of the radioactive waste are politically hard to solve. And the signatory states of the Kyoto Protocol have only now agreed on a reduction of CO_2 emissions. This, however, rules out a massive increase in the use of coal to produce hydrogen,

Renewable Energy. Edited by R. Wengenmayr, Th. Bührke. Copyright © 2008 WILEY-VCH Verlag GmbH & Co. KGaA, Weinheim. ISBN 978-3-527-40804-7

TAB. 1 | WHICH PROBLEMS COULD HYDROGEN SOLVE?

Problem	Solution with H_2?
Increasing scarcity of energy raw materials	no
Climate protection	no
Motor fuel for transportation	yes
Power grid stability	maybe
Indirect electric power transport	no
Alternative to storage batteries	yes
Decentral heat-power-production	no
These statements hold for hydrogen generated by electrolysis	

as long as there is no effective technology for the sustainable separation and storage of this greenhouse gas – and such a technology is rather improbable within the next 25 years.

In regard to this question, the USA has adopted a quite different political viewpoint from that of Europe: The current US government has denied for a long time any causal connection between the emission of greenhouse gases and the climate change; and even now, it is still opposing any strict and quantitative goal to reduce greenhouse gas emissions. Furthermore, it has no plans for terminating the use of nuclear power generation. Thus, its hydrogen plans are closely associated with hydrogen production via fossil fuels or nuclear power. In Europe, climate protection is seen as a political goal with the highest priority, and this implies going well beyond the commitments of the Kyoto Protocol. Since nuclear energy is utilized in only some of the European states, only the renewable energy sources remain as the basis for the production of hydrogen on a scale relevant to the energy economy.

Hydrogen itself thus contributes in a first approximation neither to energy supplies nor to climate protection. If it is produced from renewable energy sources, it is about as secure a supply and as climatically neutral as electric power from renewable energies. This assumes that recent suspicions are not confirmed, according to which hydrogen which escapes through leaks could itself contribute to the greenhouse effect in the upper atmospheric strata.

A Decisive Aspect: Fuel for Automobiles

One thus has to look at second-order effects to discover just how hydrogen could contribute to the solution of concrete problems.

The most important of these problems is the supply of motor fuel for automobile transportation. Here, electric power can be used within the travel range of a "full tank" only with considerable limitations. There is also little hope that storage batteries for electric vehicles will be greatly improved. Thus, hydrogen offers important advantages, because it can in principle store more energy on board of a vehicle (see the info box "Hydrogen Storage in Vehicles", p. 85, and pp. 96). It can be stored as pressurized gas at ca. 800 bar (1 torr), or also as liquid hydrogen at –253 °C (-423 F), and would then offer the same range of travel as the usual full tank of petroleum fuel. For the necessary conversion of hydrogen into mechanical driving energy, either an internal-combustion engine or a fuel cell with an electric motor can be used, assuming that the ongoing development of fuel cells proves to be technically and economically successful.

In vehicles, the hydrogen concept can doubtless be put into action. The process steps of compression or liquefaction and the transport of pressure bottles or cryogenic hydrogen in tank trucks would present no decisive technical problems. The losses of 10 % on compression and 25-30 % on liquefaction are not in principle alarming. However, they burden the ecological and economic balance, in which other losses must also be taken into account. Among these are for example up to 25 % loss during electrolysis. And at least in the case of cryogenic hydrogen, one must also remember that ca. 1 % of the hydrogen is lost every day from the tank; when the vehicle is not used often and the tank is thus filled at long intervals, this is a noticeable disadvantage.

Many of the application technologies of hydrogen have already been successfully demonstrated or are even routine. Hydrogen is already produced, traded and transported in large quantities today, especially for refineries. Per year, this amounts to over 150 million $t_{oil\ equivalent}$; one $t_{oil\ equivalent}$ corresponds to the amount of H_2 with the same heat content as 1 ton of oil. The decisive point is almost entirely the price at which hydrogen can be utilized – and in determining this price, the production of hydrogen fuel and its delivery to the pump are crucial.

An additional problem, for which a hydrogen scenario is often suggested as a possible solution, is the fact that Nature does not permit uniform or demand-oriented electric power production from wind and solar plants. If, for reasons of climate protection, renewable energies are to take over a large portion of the electric power generation in the future, this unreliability could cause problems with the voltage and frequency stability of the power grid.

To solve these problems, strategies for influencing demand and storage technologies are required. Hydrogen could serve as such a storage medium. When the supply of power is greater than the demand, i.e. at times of low power consumption, electrolysis plants could produce hydrogen using the excess power. When power is in short supply, stored hydrogen could be converted back into electric power by gas turbines, combustion engines or fuel cells, and thus supply additional energy to fill the gap. Hydrogen is here in competition with other storage technologies such as compressed air in underground caverns. In Alabama (USA) and Huntorf, in Northern Germany, such a system is already in operation: A salt cavern with a volume of 310,000 m^3 (1.1 mio cu ft) at a depth of 650 to 800 m (2,100-2,600 ft) and at 60 bar (0.1 torr) pressure is used as a large compressed air tank; this corresponds to the energy required for two hours of full operation at 290 MW.

The possible use of hydrogen storage systems for the power grid should however not be regarded as the basis of a motor-fuel supply for transportation. The amount of hydrogen which would have to be stored, corresponding to about 10 % of the annual electric power production of a supply area, is much too small in comparison to the requirements for transportation: these are around 100 % of the annual electric power production, if all the motor fuel were to be obtained from electrolysis. As a result, the annual electric power production would have to be doubled in order to supply both the electricity and the motor-fuel requirements – assuming that major improvements in efficiency and conservation do not noticeably reduce the demand.

On the other hand, an electrolysis plant also produces oxygen, since it dissociates water molecules. It would therefore be advantageous to utilize both gases locally for generating electric power. Conversely, an interplay of both applications, for stationary electric power generation and for transportation, might result. If indeed electrolysis hydrogen were to be produced in large amounts for use in transportation, the operation of the electrolysis plants could also be used to regulate the power grid: They would be taken off line at times of peak demand from the other power consumers as part of a demand side management system.

Power from the Congo River

Renewable energy sources for the production of hydrogen need not necessarily be right next door. As in the case of petroleum, it might be expedient to transport either the energy (as electric power) or hydrogen which had been generated with it to Central Europe, for example. Previously, experts held that the transmission of electric power over thousands of kilometers, e.g. from hydroelectric plants on the Congo River to Central Europe, would be so problematic that they considered hydrogen pipelines as an alternative. According to calculations, the conversion losses from the electrolysis on the Congo and the losses from power generation in the region where it was to be consumed would together be less than the losses due to a direct power transmission from Africa to the consumers.

This calculation was the basis for the prediction that it would be necessary to construct large-scale hydrogen systems. In view of the progress in high-voltage direct-current power transmission (HVDC), this prediction is no longer valid. Today, HVDC is held to be economically superior even over such long distances. The technical maturity of superconducting cables could in the long term even further improve the prospects for long-distance power transmission. However, one should note that high-voltage transmission lines cannot be arbitrarily run through densely populated areas. Resistance on the part of the affected population could force the acceptance of economic compromises. Currently, investigations are underway of the technical feasibility of transporting power in the Gigawatt range from wind parks in the North Sea to land. Hydrogen, at least in the long term, is not out of the question as an alternative, since here there are competing problems in transporting and connecting the electric power to the grid.

Portable Devices and Combined Heat-Power Production

A fourth problem area is related to the disadvantages of storing electric energy in batteries: This affects the useful life of portable devices such as mobile telephones, laptop computers, electric power tools or wheelchairs. Here, fuel cells offer the possibility of lengthening the useful operating time, once they have become sufficiently compact and inexpensive; batteries have specific weights per unit of stored energy which are about two orders of magnitude more than the value for hydrogen quoted above (see pp. 96). Such devices would be operated with pressurized hydrogen cartridges or with methanol. For the fuel cells, such applications probably offer good entry markets with learning effects for industrial production. As a hydrogen technology, however, they have no significance for the energy economy, since the amounts consumed are too small.

The use of electrolysis hydrogen from renewable energy sources in installations for small-scale combined heat-power production is, in contrast, not expedient. Such installations are used for space heating and hot water heating, and utilize part of the fuel's energy at the same time to generate electric power; they thus have an especially high efficiency. If electrolytically generated hydrogen is used as the fuel for such an installation, one is obviously using an

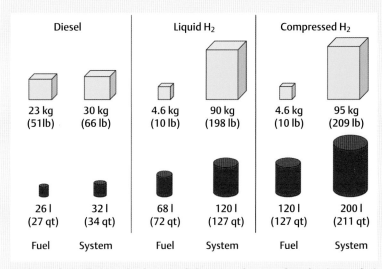

HYDROGEN STORAGE IN VEHICLES

Hydrogen indeed contains more energy per unit *mass*, at 33 kWh/kg (15 kWh/lb), than diesel fuel (ca. 14 kWh/kg (approx. 6 kWh/lb)), but not per unit *volume*: as a liquid, it contains 2.35 kWh/l (2.22 kWh/qt), as a gas at standard pressure only 0.003 kWh/l (0.003 kWh/qt). The graph also shows how much weight or volume is required in addition by the structure of the tank, which has to provide the necessary insulation and rigidity (the "system", in the right-hand column in each case). Metal-hydride storage systems, not shown here, can at present store only an amount of hydrogen which corresponds to roughly 2 % of their own weight.

Diesel		Liquid H$_2$		Compressed H$_2$	
23 kg (51lb)	30 kg (66 lb)	4.6 kg (10 lb)	90 kg (198 lb)	4.6 kg (10 lb)	95 kg (209 lb)
26 l (27 qt)	32 l (34 qt)	68 l (72 qt)	120 l (127 qt)	120 l (127 qt)	200 l (211 qt)
Fuel	System	Fuel	System	Fuel	System

Tank weight (yellow) and volume (red) for a travel range of 400 km (248 miles) in each case. (Graphics: Andrea Ballouk, FZ Jülich.)

indirect electric heating system which is made more expensive by conversion losses. This conclusion is independent of whether the conversion device is a conventional internal-combustion engine, a Stirling motor, a fuel cell or a gas microturbine. The conversion of electric power generated by a primary energy source into hydrogen, with accompanying costs and losses, only to use it to again produce heat and electric power, is more reminiscent of a public joke than a vision of an efficient energy technology for the future.

Hydrogen Production

Of critical importance to the introduction of hydrogen as a fuel on a scale relevant to the energy economy is the price and cost situation. It is thus decisive how hydrogen would be produced technically on a large scale. The most probable scenario from today's standpoint is production by electrolysis, as we shall see. In this standard scenario, hydrogen

will never be cheaper than the electric power used to produce it, which in turn for reasons of climate protection must be generated without releasing CO_2. In addition, electrolysis apparatus have energy losses of ca. 25 % in practical operation – the reasons for this are mainly technical and not based on thermodynamics.

The most abundantly available renewable energy sources are the Sun and wind energy. Photovoltaics (PV) (see pp. 34 and pp. 42) and wind power plants (pp. 14) can generate electric power directly from them. They are thus the main candidates for the production of "renewable hydrogen", and they practically demand the standard scenario of electrolysis. However, if one also takes into account the fact that PV and wind installations or solar-thermal power plants (see pp. 26) will, in the medium term, not be sufficient to supply the remaining demands for electric power, then realistically, there is no capacity from renewable energy sources for the production of hydrogen in sight: It is not reasonable to tap renewable electric power from the grid as long as this power must be replaced by burning more fossil fuels. In that case, producing hydrogen for transportation would simply reduce the potential of renewable energies for climate protection, and worsen their economic position.

It is in any case logical that hydrogen will be generated using electric power only when natural gas becomes more expensive on the energy market than electricity. At the moment, power generation using natural gas is indeed still on the rise. Our hydrogen scenario will only become realistic when this type of utilization of natural gas has definitely come to an end. Fundamentally, natural gas – in terms of its current use as a gaseous energy carrier – is indeed to a large extent equivalent to hydrogen (see the infobox "Thermodynamics"). As far as costs are concerned, we know for example from the legally defined feed-in tariffs for renewable power into the grid in Germany that they are still rather

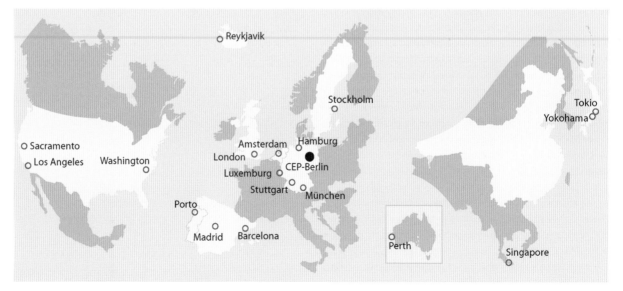

Overview of the current international hydrogen projects. (Graphics: CEP.)

high; hydrogen produced with this power would not be marketable in the foreseeable future.

Of course, research is underway on how to make electrolysis more efficient. One idea is the development of high-temperature vapor-phase electrolysis. In principle, this would be the reverse of the processes in a high-temperature fuel cell with ceramic-oxide electrolytes: Such a cell would decompose hot water vapor into H_2 und O_2 by means of an electric current. It could be operated at a lower cell voltage – and thus exergetically more favorably – than the standard electrolysis. But this process, like every other form of electrolysis, depends on the price of electric power. Furthermore, it also requires an inexpensive source of high temperature heat produced in a CO_2-free manner.

For the standard scenario of electrolysis hydrogen from renewable energies, we can thus draw the following preliminary conclusions: It will only then become economically realistic, when natural gas and petroleum have been forced out of the power-generating market due to their scarcity and high costs. Currently, there are speculations that the worldwide maximum for natural gas and oil production will be reached within the coming two decades, and this will give rise to corresponding price increases. On the other hand, the development of the renewable energy sources worldwide is proceeding only very slowly, in contrast to the situation in Germany. This is due to the fact that among other things, their costs are still rather high, in particular those of photovoltaic energy conversion. Economically competitive "renewable" hydrogen is thus hardly to be expected within the coming two to three decades.

Now there are in principle various other possibilities for producing hydrogen without CO_2 emissions in addition to electrolysis. However, these developments are still in the basic research stage. For several decades, for example, the idea of thermochemically decomposing water (see the infobox "Thermochemical Water Dissociation") has been discussed. It employs a catalytic process at high temperatures. The necessary thermal energy (at ca. 1,000 °C (1,830 F)) could be produced by using solar radiation: As in a solar-thermal power plant, mirrors could concentrate the Sun's rays so that at the focus point, sufficiently high temperatures for the desired reaction would be obtained. Whether such a thermochemical cycle could achieve a better efficiency and economic competitiveness – in spite of the considerable materials problems due to the high temperatures – than solar-thermal electric power generation followed by electrolysis, is still an open question. It is up to research to provide an answer.

From Algae to the Biomass

There are also biological pathways by which algae or other organisms can produce hydrogen directly from water using sunlight. They are to be sure still the objects of basic research. Efficient technical solutions and proven statements concerning economic feasibility cannot be expected for some time to come.

The biomass can in principle also serve as a source of hydrogen, i.e. organic rubbish, agricultural wastes or even intentionally cultivated energy-yield plants. There are various routes to this end, for example via biogas from moist biomass, or gasification of solid biomass such as wood or straw. Work is in progress on long process chains, e.g. for the production of hydrogen from alcohols, which of course must first be extracted from the biomass. Fundamentally, there are two problems: On the one hand, inexpensive biomass is available only in limited quantities; and secondly, the production of heat and electric power by combustion of the biomass or biogases is the more attractive route to energy recovery in terms of the environment, climate protection, and economic factors. All this argues against a hydrogen infrastructure which makes use of biomass as its primary energy source.

Thus far, we have ignored the method of hydrogen production which is currently the most prevalent: reforming of natural gas. The reason is of course that this production path is not compatible with the assumed scenario of increasing scarcity of natural gas and consistent climate protection. At the same time, the use of reformer hydrogen could be expedient during a transition period for introducing the technology, in particular if fuel cells can compensate the conversion losses through their high efficiencies. Fuel cells have the advantage that the electrochemical conversion does not go by the route of hot combustion gases. Therefore, their efficiencies are not limited by Carnot losses as required by the Second Law of Thermodynamics.

Can CO_2 Separation Alter the Situation?

These considerations will also include the question of whether there can be a climatically neutral use of fossil-fu-

---

THERMOCHEMICAL WATER DISSOCIATION

Above 4,500 °C (8,100 F), water dissociates spontaneously into H_2 und O_2. At a technically maneageable temperature around 1,000 °C (1,800 F), various thermochemical cyclic processes can be used to dissociate water. An example which is being considered in the USA is the sulfur-iodine cycle, which is soon to be demonstrated in Idaho using process heat from a high-temperature nuclear reactor. It involves the following reaction steps:

1. at 120 °C (248 F), exothermic:

$$I_2 + SO_2 + H_2O \rightarrow 2HI + H_2SO_4,$$

2. at about 850 °C (1,560 F), endothermic:

$$H_2SO_4 \rightarrow SO_2 + H_2O + 1/2\ O_2,$$

3. at about 350 °C (660 F), endothermic:

$$2HI \rightarrow I_2 + H_2.$$

el energy sources, in particular with coal, which is available in large quantities, by means of CO_2 separation and long-term storage. Work is being carried out on such a sequestration, but a responsible technology is not likely to be mature before 2025. It is also unclear whether such a technology would lead to cheaper energy than the renewable energy sources.

Provided that such a carbon capture and sequestration process (CCS) can be established, hydrogen could also be produced by gasification of coal with steam. The product of the first reaction step, synthesis gas which still contains CO, could be shifted in its chemical composition towards H_2 and CO_2 and then separated into H_2 and CO_2. Thereafter, the hydrogen could be used in a combined gas and steam turbine cycle to generate electric power, or in purified form as hydrogen fuel for vehicles powered by fuel cells.

If such a CCS scenario some of the statements made above concerning our standard scenario of electrolysis using electric power from renewable sources would have to be revised. In this case, for example a decentralized combined heat-power production using hydrogen and fuel cells would again become reasonable, since it would no longer represent an indirect form of electric heating.

Conclusions and Outlook

The 21st century will not experience a hydrogen society or a hydrogen economy. If it makes sense at all to name a society or an economic system after a secondary energy carrier, then electric power would be the appropriate eponym. Electricity is an ideal secondary energy carrier for nearly all applications, even some modes of transportation like trains, trams and trolleys. Only for mass transportation on land, water and in the air it is less suitable, owing to the limited storage capacities of batteries.

But the system of electric power supply and consumption requires effective storage methods, if it is to be based essentially on renewable energy sources. Here, hydrogen comes to the fore, especially in transport. In the foreseeable future, however, all of the CO_2-free energy sources must be used to displace coal and hydrocarbons from the heat and electric power markets, insofar as this is economically feasible. Thus for a long time to come, there will be no free capacity on the electric power market for the CO_2-free production of hydrogen on a large scale. If, for example, through improved insulation and energy efficiency we are able to successfully banish oil and gas from space heating of houses and buildings, then we can continue for some time to fill the tanks of our – hopefully clean and thrifty – automobiles with petrol or diesel fuel.

Summary

Hydrogen, like electric current, is a secondary energy carrier, which must be produced using a primary energy source. For its CO_2-free, industrial-scale production, electrolysis is the process of choice; other environmentally-friendly methods are not yet technically mature. Such a hydrogen economy is reasonable as a protective measure for the climate only if the electric current used for the hydrogen electrolysis comes from a CO_2-free energy source. If nuclear energy is undesirable, then solar or wind-generated power is most suitable, since the Sun and the wind are available in sufficient quantities worldwide. The cost of generating this primary energy thus determines the price of the hydrogen produced. Hydrogen is comparable to natural gas as an energy carrier. It will have a chance to become economically competitive only when natural gas becomes more scarce and more expensive per kilowatt hour than electrical energy. In comparison to electric power, hydrogen has an advantage only in the area of mobile applications for vehicles and for mobile machinery: this is due to its superior energy storage capacity. Efficient and economically competitive fuel cells could further increase this advantage.

About the Author

Gerd Eisenbeiß studied physics at the University of Karlsruhe and received his doctorate in process technology there. From 1973 to 1990, he worked in various capacities for the German Federal Chancellor's Office and the Federal Research Ministry. Thereafter, he was for 11 years Program Director for Energy and Transport Research in the German Aerospace Center, and from 2001 to 2006, he was Director at the Jülich Research Center (Jülich, Germany), responsible for energy and materials research. He retired in 2006.

Contact:
Dr. Gerd Eisenbeiß,
gerd.eisenbeiss@t-online.de

Seasonal Storage of Thermal Energy
Heat on Call

BY SILKE KÖHLER | FRANK KABUS | ERNST HUENGES

In energy-supply systems with combined heat and power (CHP) generation, or also in combination with renewable energy sources, especially solar energy, the question often arises as to how to store the thermal energy. This can involve short storage times such as hours or days, but it continues on up to annual (seasonal) heat storage.

Solar radiation and wind are – as everyone knows – not available around the clock; instead, they occur at variable times and with limited predictability. Only through the use of energy storage systems can these renewable energy sources be made reliable and readily usable for energy supply. Electrical energy can, for example, be converted into chemical energy – the catchword here is hydrogen economy (see p. 83) – and thus stored. In the case of thermal energy, in contrast, the seasonal variations make it attractive to preserve the excess heat from the summer for use during the winter, or conversely, to use the cold of winter for cooling in the following summer.

The same is true of installations for combined heat and power production (CHP). They make use of the generated heat as well as the electrical energy. Thus, efficiencies of CHP systems are very high. The deployment of such installations has to be oriented towards the end-user energy requirements, i.e. electrical power or heat. Accordingly, one refers to heat-operated (the determining parameter is the requirement for heat) or power-operated (determining parameter is the requirement for electrical power) CHP installations.

Most installations are heat-operated, since the electrical power not needed by the local user can always be fed into the power grid, which acts as a large storage system. Since, however, the price of electrical energy is much higher than that of heat energy, it would often be preferable from the economic standpoint to employ power-operated systems. The operator could then compensate expensive peak loads by means of in-house production, for example. If the co-generated heat shall not be released as waste heat to the environment, it has to be stored. This excess heat can then be stored in a seasonal reservoir. Storage of excess thermal energy is thus finally the precondition for energetically and economically expedient power-operated CHP installations.

The Thermodynamics of Energy Storage

Such a storage system consists essentially of four functional units: the storage medium, the charging and discharging systems including the heat-transport medium, the storage container structure, and ancillary systems.

In order to classify systematically the types of storage, the well-known distinction between open and closed systems, defined in thermodynamics, is useful here. In open systems, energy and matter can be transported across the system boundary. In closed, non-adiabatic systems, only energy and not matter can be transported across the system boundary.

The corresponding artificial or geological structures are popularly referred to as heat or cold reservoirs. This is problematic, since heat is always just a transport quantity appearing when passing across a system boundary, and in this sense cannot be stored. Like work, it is a process quantity and is therefore dependent on the path of a process. The heat transported across the boundary of the storage system changes the internal energy, which in most cases is reflected in a change of the temperature of the storage medium. It would thus be more physically correct to speak of the storage of thermal energy.

Whether a storage system serves as a source of heating or of cooling in an energy-supply installation is in the first instance unimportant for the construction of the system. Thus, some thermal storage systems are used in alternate modes, and supply cooling in summer and heating in winter. To be consistent, we will refer in the following to charging of the storage system when heat is input to it, and to discharging when heat is extracted from the storage system. Since these terms have become common, we will continue to refer to heat reservoirs (when the system is mainly used as a source of heat) and cooling reservoirs (when it mainly serves as a source of cooling or as a heat sink).

Considering the energy balance with the surroundings of a reservoir, its interactions with these surroundings and its losses can be specified. The system boundary contains the storage medium, the charging and discharging system, and the heat-transport medium. The sum of the heat input and output and the heat losses equals the change in the internal energy of the storage system. The quantity of heat input or output is equal to the temperature change of the storage medium multiplied by its heat capacity, i.e. the product of its volume, density and specific heat.

Owing to the temperature change of the storage medium, there is a temperature difference relative to its sur-

Renewable Energy. Edited by R. Wengenmayr, Th. Bührke. Copyright © 2008 WILEY-VCH Verlag GmbH & Co. KGaA, Weinheim. ISBN 978-3-527-40804-7

FIG. 1 | AN AQUIFER STORAGE SYSTEM

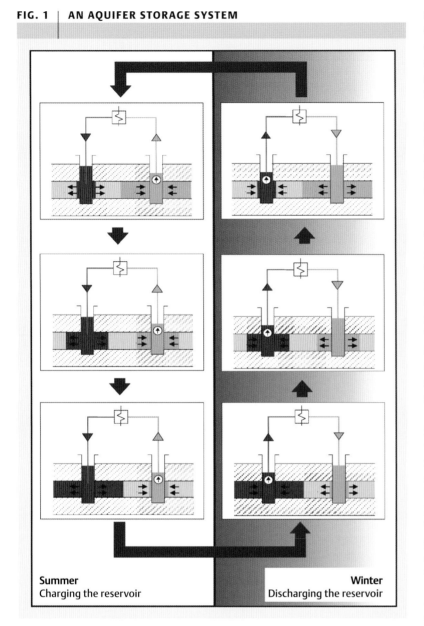

Summer
Charging the reservoir

Winter
Discharging the reservoir

Charging and discharging of an aquifer storage system (Source: GTN).

system across its boundary (charging and discharging plus exergy losses accompanying heat losses), in addition to a conversion term. The latter takes into account the exergy destruction, which for example occurs on mixing within stratified hot-water storage systems by convection, or through thermal conductivity.

A goal in the construction of storage systems is of course to always reduce the losses as far as possible, i.e. to improve the storage efficiency. It will be seen, however, that in designing different types of seasonal storage systems, such as aquifer and hot-water reservoirs, quite different concepts are pursued.

Aquifers as Seasonal Storage Systems

The technology of aquifers as energy storage systems is in many respects closely related to hydrothermal geothermics. The storage medium is the natural substratum, i.e. the rock layers and the deep water they contain, which also serves as heat-transport medium. Aquifer storage systems are as a rule accessed via two boreholes or groups of boreholes. These are placed at a certain distance from each other, in order to avoid mutual thermal influences. The systems are open below ground and closed above ground. Above ground is a heat exchanger, so that only energy transport – and no matter transport – occurs.

Both boreholes are fitted out with pumps and injection piping, which allows the flow of the heat-transport medium in the above-ground part of the plant to pass through in either direction. For charging, that is for storing thermal energy in the reservoir, water is taken from the cooler boreholes, warmed above ground, and injected into the warm boreholes. For discharging, the direction of flow is reversed: The pump in the warm borehole extracts water up to ground level, where it transfers heat to the energy supply system. The thermal power transferred is proportional to the mass flow rate of the thermal water in each case.

Charging and discharging of the aquifer storage system takes place horizontally (Figure 1). Around the warm borehole, a heated volume of water forms, which is pumped out again on discharging. Since the thermal water serves both as heat-transport medium and as storage medium, the temperature present in the reservoir is also available above ground, as long as the thermal losses within the borehole can be neglected. The discharging temperature will, however, always be lower than the charging temperature, since the warm water volume cools at its outer boundary. This heat transport to the cooler surroundings takes place through heat conduction and natural convection, since the warm water mixes with the cooler water from the surroundings, and through groundwater flows.

From this, we can derive the initial requirements for aquifers which are to be used for seasonal energy storage:
• The geological formation should be closed above and below, and the natural flow rate of the ground water as low as possible (preferably zero), to prevent the warmed or cooled water from flowing away.

roundings. This temperature difference leads to storage losses. A closed storage system for example loses thermal energy via heat conductivity across the container walls. In an open storage system, heat can also be transported in a mass flow across the boundary of the system.

Along with such external storage losses through interactions with the surroundings, internal storage losses must also be considered. These occur as a result of equilibration processes within the storage system, such as a flow between the layers in a stratified hot-water reservoir. These losses can be assessed using the exergy balance around the reservoir. The exergy is that portion of the energy which can be unrestrictedly converted into any other form of energy. The change in the exergy in a storage system is the sum of the exergies input into the system and output by the

- As a rule, the temperature below ground increases with increasing depth. Heat storage reservoirs will thus be more readily found in deep strata than cooling reservoirs, since then the natural temperature of the aquifer lies nearer to the desired mean storage temperature, and storage losses are reduced. The requirement that the natural aquifer temperature be in the range of the mean storage temperature cannot always be fulfilled in an economically feasible way, in particular for high-temperature storage. A rough calculation shows: In order to obtain a temperature of 70 °C (158 F) below ground, given an average temperature increase with depth of 30 K (54 F) per kilometer, and 10 °C (50 F) external temperature, boreholes of two kilometers (1.2 mi) depth would be required.
- For a desirably low pumping power, a high water throughput in the horizontal direction within the rock layers is needed.

The maximum storage temperature and the storage volume, which determine the total storage capacity, are given by the natural properties of the aquifer. An advantage of aquifer storage is the relatively low investment cost. It is about 25 €/m³ (1 $/cu ft) (including planning, excluding taxes, for storage volumes of more than 100,000 m³ (3,570,000 cu ft)).

Aquifer Reservoirs for Energy Supply – Reichstag and Neubrandenburg

In the construction and the energy supply of the new government buildings along the Spree River in Berlin, forward-looking, environmentally responsible, and exemplary energy concepts were required. The energy supply system of the Berlin parliament buildings therefore contains, along with components for combined heating, cooling, and power generation, two aquifer storage reservoirs (Figure 2). The heat

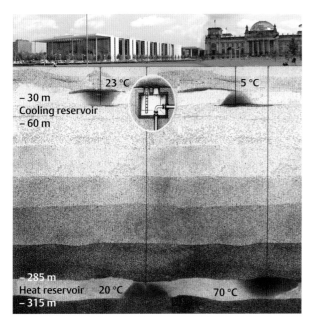

Fig. 2 An aquifer reservoir for heating and cooling (Source: GTN, BBG).

FIG. 3 | INPUT AND OUTPUT RESERVOIR TEMPERATURES

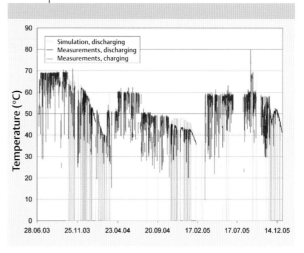

Charging and discharging temperatures of the thermal storage medium. Red and dark blue lines: measurements; light blue: model calculations.

reservoir increases the contribution of the combined heat and power plant to the overall energy supply, even when the system is power-operated. The cooling reservoir permits the use of the low winter temperatures for air conditioning in the summer months.

The storage region is a sandstone stratum at a depth of 285 to 315 m (935 – 1,033 ft), which contains salty water (brine). Two boreholes access this aquifer. Usually, it is charged at a maximum of 70 °C (158 F), and discharged at 65 – 30 °C (149 – 86 F). In operation, a volume flow rate of up to 100 m³/h (3,570 cu ft/h) can be extracted or injected. The maximum charging power is about 4.4 MW. Heat from the discharge supplies the low-temperature sectors of the various building heating systems via direct heat exchange. Additional cooling of the water (down to a minimum of 20 °C (68 F)) can be carried out using absorption heat pumps as needed; their installed cooling power is approximately 2 MW.

The project is one of a kind and is accompanied by extensive scientific monitoring. Among other things, numerical models of the substrata have been constructed within the framework of research projects; they allow the temperature evolution in the storage media to be predicted. Figure 3 shows as an example the temperature at the borehole head of the warm borehole during the time period from June 2003 to December 2005. This period includes nearly three charging and discharging cycles. The decreasing output temperature during discharging is characteristic of aquifer storage. It results from the thermal losses described above, due to thermal conduction and convection at the perimeter of the warm storage volume.

At a notably smaller depth of 50 m (164 ft), under the bend of the Spree, another aquifer storage system was developed. It is used mainly for cooling of the buildings. In operation, non-salty ground water is cooled in winter to 5 °C (41 F). This is carried out essentially at outside temperatures below 0 °C (32 F) in dry cooling towers by heat exchange with the cold surroundings. In the summer, this cooled reservoir supports the air conditioning systems via direct heat exchange.

In Neubrandenburg, a city in Mecklenburg-Vorpommern, a major portion of the buildings are connected to a central 200-MW district heating network. Its base load is carried by a combined cycle gas turbine power plant with 77 MW of electrical output power and 90 MW thermal power. Since heat supplies in Neubrandenburg in the summer are essentially used only for providing warm water, the thermal power consumption is relatively low. As a rule, it is considerably less than the heat, which is produced even when electrical power is being generated at minimum load. The difference of up to 20 MW was in the past simply released to the environment through a re-cooling plant. Today, part of this 'waste heat' is fed into an aquifer storage system and used for heating in the winter in a portion of the district heating network which operates at a relatively low temperature level. The thermal power in this network is 12 MW at 80 °C (176 F) input temperature and 45 °C (113 F) return temperature.

The aquifer storage system consists of a cool and a warm borehole spaced approx. 1,300 m (4,260 ft) apart. Both boreholes tap the operating stratum at a depth of about 1,200 m (4,940 ft), and they can extract or inject 100 m³/h (3,570 cu ft) of thermal water. The natural temperature of the stratum at this depth is about 55 °C (130 F).

In summer, cooler water is pumped out at 40 °C to 50 °C (104 – 122 F) and is stored underground at a temperature of 80 °C (176 F). In winter, the flow direction of the storage cycle is reversed; water is now extracted from the warm borehole. The output temperatures lie between 80 °C and 65 °C (176 – 149 F), depending on the time the water is extracted.

FIG. 4 | EXCESS HEAT

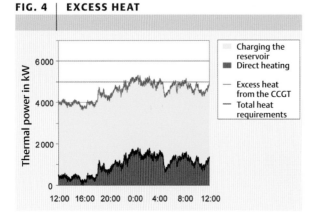

Excess heat and its use on the 5th and 6th of November 2005 in the Neubrandenburg aquifer reservoir. b) Temperatures in the district heating network and the storage reservoir on the same days.

FIG. 5 | GROUND COUPLED HEAT EXCHANGERS

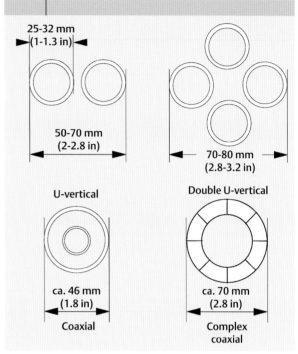

Cross sections of typical ground-coupled heat exchangers.

According to plan, it is expected that in the months April to September, a quantity of thermal energy equal to 12,000 MWh can be stored in the reservoir. In the winter, 8,800 MWh of this shall be recovered through direct heat exchange at a thermal power of 4.0 to 2.9 MW.

The Neubrandenburg aquifer storage system is now operating in its second regular annual cycle. Figure 4 shows as an example the operating regime of the reservoir during charging.

Storage in a Borehole Field

For the storage of thermal energy in borehole fields, the underground strata serve as the storage medium. Here, the reservoir is tapped by a number of boreholes of 20 to 100 m (66 – 328 ft) depth, usually symmetrically arrayed, and designed as ground-coupled heat exchangers.

A geothermal borehole field is an underground closed system in which there is no hydraulic connection between the storage medium (earth) and the heat transport medium. The heat transport for charging and discharging takes place through the walls of the tubing. The charging and discharging thermal power is thus proportional to the surface area of the tubing and to the temperature difference between the heat transport medium and the storage medium. The cross-sectional area of the tubing is therefore dimensioned in such a way that the probe surface area is as large as possible.

Figure 5 shows four types of common tubing in cross section. The U- and double U-vertical tubing comprises two or four tubes, respectively, which are connected at the bottom of the borehole. In coaxial tubes, the medium flows downwards in the outer space and upwards through the inner tube. A water-glycol mixture is used as heat-transport medium, providing sufficient antifreezing protection at low temperatures.

Owing to the temperature difference between the heat-transport medium and the storage material, which is required for heat transfer, the temperature in the reservoir is always higher (when heat is being extracted from the reservoir) or lower (during charging) than the temperature of the heat-transport medium.

These losses appear in the storage balance as exergy destruction. The heat-transport medium in a borehole-field storage system has a smaller temperature range in comparison to that in an aquifer storage system. Therefore, in energy-management systems with borehole-field storage, heat pumps, which raise the heat to a usable temperature level, are usually employed.

The heat-transport medium flows through the heat-exchanger tubes and, during winter operation, takes on heat from the surrounding earth. In summer operation, it gives up heat to the earth and thus recharges the reservoir. In contrast to aquifer storage systems, the heat-transport medium always flows through the boreholes in the same direction. Temperature variations and the storage capacity of the borehole field are dependent on the composition of the subsurface earth and on the heat-transfer properties of the tubes and their thermal contact with the earth. The reservoir cannot be isolated from its surroundings below ground. In order that the stored thermal energy not be lost, such reservoirs are usually located at sites where the ground water has only a very low – or zero – flow velocity.

Example: The Max-Planck Campus in Golm

In Golm, near Potsdam, in 1999 the Max Planck Society (a German non-profit research organization) founded a new science campus, comprising three institutes. The energy management system is based on a combined heating-cooling-power system, making use of a geothermal borehole field (Figure 6). This concept includes a block heat and power plant driven by an engine; the excess heat is used for space heating and warm water.

The borehole field contributes to cooling in the summer months, and in the process, its temperature rises. In the winter, it serves as a heat source providing heat energy for a heat pump, and is thereby cooled, causing the temperature in the borehole field to sink below that of the surrounding earth. The field consists of 160 boreholes, each with a depth of 105 m (345 ft), and it occupies an area of 65×50 m (213×164 ft), with an earth volume of about 400,000 m^3 (1,300,000 cu ft). Its overall storage capacity is 2.24 MWh, and the input/output power is nominally 538 kW.

The temperature is measured year-round at four of the ground-coupled heat exchangers. Three of these are within the field, while the fourth gives values from the undisturbed region outside the field for comparison. Each monitored borehole contains four temperature sensors at depths of 15 m (50 ft), 40 m (130 ft), 70 m (230 ft), and 100 m (330 ft), respectively. The measured temperatures during the time period from September 2001 to September 2002 in one of the boreholes within the field are shown in Figure 7. Normally, the natural temperature increases with increasing depth. During charging, the warm heat-transport medium flows downwards and thereby cools; thus, the temperature gradient in the upper part of the borehole may be reversed after charging. In the lower part of the borehole, it maintains its original sense at all times.

Hot-Water Storage

Hot-water storage systems are the most widespread type of reservoirs for thermal energy. They are used for example in

Fig. 6 *Borehole field beneath the Max-Planck Campus in Golm* (Source: H. Jung).

FIG. 7 | THE BOREHOLE FIELD IN GOLM

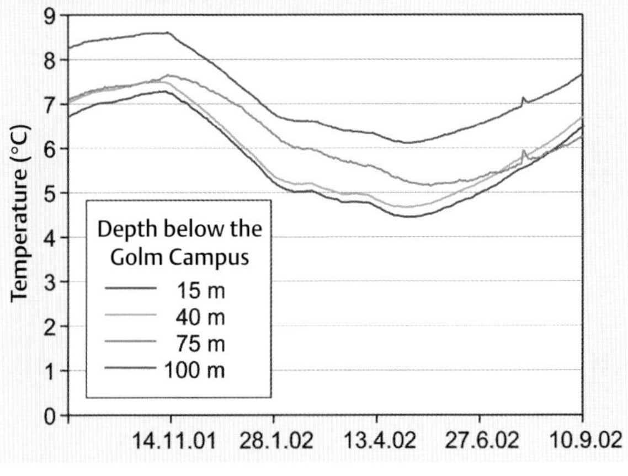

Temperature (°C)

Depth below the Golm Campus
— 15 m
— 40 m
— 75 m
— 100 m

14.11.01 28.1.02 13.4.02 27.6.02 10.9.02

Measured temperatures from September 2001 to September 2002 within the bore-hole field of the Max-Planck Campus at a depth of 15 m, 40 m, 70 m and 100 m.

FIG. 8 | A HOT-WATER STORAGE SYSTEM

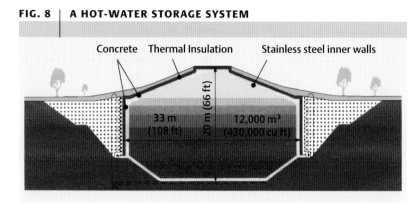

Concrete   Thermal Insulation   Stainless steel inner walls

33 m
(108 ft)

20 m (66 ft)

12,000 m³
(430,000 cu ft)

Schematic of the hot-water reservoir in Friedrichshafen (Graphics: ilek, Univ. Stuttgart).

solar-thermal plants for short-term intermediate storage. The reservoir is filled with hot water which is heated (charging) or cooled (discharging) in a heat exchanger. The storage medium remains within the system in normal operation; only heat is transferred across the system boundary. These reservoirs thus represent closed, non-adiabatic systems with respect to their surroundings.

However, hot-water storage systems are operated not only with pure water. Aqueous solutions, such as brines, are used, and even additional storage media such as gravel can be employed. This generalisation allows a better classification of the typical structural forms, tanks and basins. A common characteristic is that they are always entirely artificially constructed.

Hot-water storage systems in the form of tanks are to be found in all sizes in energy supply installations, from short-term daily storage up to seasonal storage systems. For the latter, usually large cylindrical containers are employed, which can either be buried underground or set up above ground. Storage reservoirs at ground level or above must be able to withstand the environmental conditions of their surroundings and be adapted to them. Their walls are usually fabricated of reinforced concrete and are thermally insulated from their surroundings. Thermal losses occur only through heat conduction to the surroundings.

Water serves as the heat-transport and the storage medium. Due to this identity, the storage temperature is made available to the energy system as the usable operating temperature at its output, as in aquifer storage systems. Within the reservoir, the temperature decreases from above to below. On discharging, the water with the highest temperature at the uppermost point of the reservoir is extracted, cooled in the energy-transfer system, and then returned to the lowest, coolest region in the reservoir. This temperature stratification, which normally will already appear as a result of density differences, will be affected by thermal conduction and convection. While thermal conduction within the reservoir cannot be avoided, convection can be held to a minimum. Thus, for example, mechanisms for storing the thermal energy within the layer of the same temperature

can be useful. This is especially interesting for solar heat, which can be stored at varying temperatures.

Since the storage reservoirs are usually operated at ambient pressure, the maximum storage temperature lies below 100 °C (212 F). In contrast to aquifer storage systems, they have a clearly defined and limited storage capacity. The investment costs of these storage systems are specified to be in the range of 450 – 120 €/m³ (18 - 5 $/cu ft)(depending on the overall volume) for concrete reservoirs, and 3,000 – 600 €/m³ (120 - 24 $/cu ft) (for 0.2 – 100 m³ volume (7 - 5,570 cu ft)), or below 100 €/m³ (4 $/cu ft) (at volumes larger than 10,000 m³ (375,000 cu ft)) for steel reservoirs.

Solar-Assisted Local Networks

The German Federal ministries for Commerce and Technology and for the Environment, Nature Protection and Nuclear Safety have subsidised the construction of pilot and demonstration plants for local solar heating networks within the framework of the German program 'Solarthermie2000'. In three of the subsidised pilot plants, in Friedrichshafen, Hamburg, and Hanover, seasonal hot-water reservoirs are integrated into the energy supply system as tank structures. In these solar-assisted local heating networks, use is made of the stored energy to preheat the return flow in the local heating network. If the heat output power or the temperature from the storage reservoir are not sufficient, heat from fossil fuels (gas, oil) or district heating grids is used to reach the desired temperature.

The largest of these hot-water storage reservoirs is in Friedrichshafen. It has a height of 20 m (66 ft) and an inner diameter of 32 m (105 ft), and thus provides a storage volume of 12,000 m³ (430,000 cu ft). In the final construction stage, this local heating network will supply heat to 570 dwellings with a heated floor space of nearly 40,000 m² (430,000 sq ft) (Figure 8).

The experience with the pilot projects for solar-assisted local heating with long-term thermal storage has thus far yielded a fraction of solar heat in the overall energy consumption of 30 to 35 %. With additional improvements to the systems technology, solar fractions of 50 to 60 % are expected.

Gravel-Water Storage Systems

Gravel-water storages are likewise artificial structures. In them, a mixture of gravel and water serves as storage medium, with a gravel portion of 60 – 70 vol. %. This mixture is placed into a cavity in the ground which has been lined with a watertight plastic sheet. The maximum storage temperature is limited by the temperature stability of the plastic, and is typically in the range of 80 °C (176 F). Due to the lower specific heat of the gravel, a gravel-water storage system requires about 50 % more volume than a pure water reservoir for the same storage capacity.

Charging and discharging can be accomplished either by direct flow through the reservoir or by heat exchange using coiled tubes. The heat-transport medium is thus either the water within the reservoir, or a second medium such as brine or an antifreeze mixture. In gravel-water storage systems, a vertical temperature stratification is likewise observed, and it can be enhanced by the charging and discharging processes.

The gravel within the reservoir has two advantages: It supports part of the static load, and thus makes the reservoir construction lighter and simpler. In addition, it reduces the free convection of the fluid within the reservoir and thereby the internal losses. The external losses are limited by thermal insulation applied outside the watertight sheet. In the solar-assisted local heating networks in Steinfurth and Chemnitz, which were also subsidised by the support programme mentioned above, gravel-water storage systems are utilized.

Summary

Energy from renewable sources is often not continuously available. The storage of thermal energy is therefore of great importance. Which type of storage technology is chosen for a particular application depends on various conditions. Seasonal storage can be applied successfully using structures such as hot-water storage in tanks and in gravel-water reservoirs. For some time, underground storage has also been practiced, using borehole fields or aquifers. The natural substratum is a complex storage medium, which requires a considerable effort for successful operation. However, such storage systems can be developed with a considerably lower specific investment cost than hot-water storage systems.

About the Authors

Dr.-Ing. Silke Köhler studied energy and process technology at the Technical University in Berlin, and from 1996 systems technology for solar plants at the Institute for Solar Energy Research in Hameln Emmerthal. Since 2000, she has been at the Geothermal Technology Division of the GeoResearchCentre (GeoForschungsZentrum) in Potsdam. She obtained her doctorate in 2005 at the TU Berlin on geo-thermally driven power generation processes.

Dr.-Ing. Frank Kabus studied thermal and hydraulic mechanical engineering at the TU Dresden; he obtained his doctorate at the TU Dresden on working materials for cyclic heat pumps. Since 1987, he has worked in the area of geothermal energy supplies. He is the general manager of Geothermie Neubrandenburg GmbH (GTN).

Dr. Ernst Huenges, physicist and process engineer, is leader of the Geothermal Technology Division at the GeoResearchCentre (GeoForschungsZentrum) in Potsdam. Currently, he is chairman of the German Helmholtz Centres for Geothermal Technology. He has participated in many deep- bore projects, for example the German continental deep-bore programme.

Contact:
Dr.-Ing. Silke Köhler, Dr. rer.nat Ernst Huenges, GeoForschungsZentrum Potsdam, Sektion 5.2 Geothermie, Telegrafenberg, D-14473 Potsdam, Germany.
skoe@gfz-potsdam.de, huenges@gfz-potsdam.de.

Dr.-Ing. Frank Kabus, Geothermie Neubrandenburg GmbH, Seestraße 7A, D-17033 Neubrandenburg, Germany.
gtn@gtn-online.de

Fuel Cells for Mobile and Stationary Applications

Taming the Flame

BY HARALD LANDES | MANFRED WAIDHAS

Fuel cells allow clean and resource-efficient energy conversion. Nevertheless, this technology has developed only slowly since its introduction in the year 1839 by Sir William Grove. A great deal of public attention has been attracted by the interest of the automobile industry. This is now changing the situation.

S everal well-known car manufactures have demonstrated the technical feasibility of fuel-cell powered vehicles in the last ten years, and have thereby attracted the attention of the mass media. The driving forces behind this return to the fuel cell are in particular the increasing scarcity of fossil energy resources, and the rise of carbon dioxide concentration in the atmosphere. Fuel cell powered drives consume less fuel per kilowatt hour or kilometer driven and release correspondingly less carbon dioxide, owing to their high electrical efficiencies. Other environmentally harmful trace gases such as nitrogen oxides can also be reduced or entirely avoided by the use of fuel cells.

If fuel cells are operated on pure hydrogen, the only "waste product" they release to the environment is water. However, since hydrogen is not available everywhere, and its storage is difficult, attempts are underway to adapt fuel cells to be operated with more readily manageable fuels such as methanol, petrol or natural gas. The operating principles and properties of various types of fuel cells are described separately in the infoboxes "How do Fuel Cells Work?" on p. 100 and "Types of Fuel Cells" on p. 101 (see also [1–3]).

Applications of Fuel Cells

Fuel cells have proven their reliability over the past decades in space applications, which also provided the stimulus for continuing technical development. The power source for a submarine drive is another application. Both are however niches which have a negligible effect on the overall energy economy in comparison to the widespread applications of fuel cells for stationary, mobile, and portable power which are currently being planned.

For stationary operation in power plants, fuel cells with power outputs in the range of 1 kW up to 10 MW are conceivable. They thus could provide decentralized electric power supply for single-family dwellings and for industrial plants as well. For these applications, most developers envisage the so-called high-temperature systems: SOFC (Solid Oxide Fuel Cell) or MCFC (Molten Carbonate Fuel Cell). For installations in the range of 10 MW, when sufficient waste heat at high temperatures is available for driving a subsequent gas and steam turbine cycles (combined cycle, CC), electrical system efficiencies of 70 % seem to be achievable. For comparison: Todays most advanced natural gas based power-plant technology, the CC plant with several 100 MW output power, attains an efficiency of "only" 58 %. An additional advantage of a fuel cell power plant would be its low trace-gas emissions (carbon monoxide, nitrogen oxides, hydrocarbons).

Road transportation represents another important anthropogenic source of CO_2 and other harmful emissions. It is responsible for about a third of the CO_2 emissions of the industrial nations. Here, electrical drives could be employed as a locally emission-free power-train concept. However, the storage of the required electrical energy is difficult. Even the most advanced battery technology allows a conventional vehicle to operate over at most 250 km (255 mi), because batteries are heavy and their storage capacity for electrical energy is still insufficient. Fuel cells, in contrast, make use of the energy which is chemically stored in fuels like hydrogen, methanol or petrol, similarly to internal-combustion engines. Their high energy content makes it in principle possible for a vehicle equipped with a fuel-cell system to exhibit a range which is competitive with that of conventional automobiles (see also the infobox "The Storage

THE STORAGE CAPACITY OF VARIOUS ENERGY STORAGE DEVICES

Equivalent to 50 l gasoline (n-octane)

7.860 l — Lead accumulator (Pb/PbO₂)

170 l
107 l
93.4 l — Hydrogen (liquid)
Magnesium hydride
50 l — Methanol
n-octane (gasoline)

TAB.	ENERGY CONTENT	
Fuel	**Energy Content**	
	per volume (kWh/l)	**per weight (kWh/kg)**
n-octane	9.43	13.33
Methanol	5.05	6.37
MgH_2	4.42	3.06
H_2 (liquid)	2.78	40.00
Lead accumulator	0.06	0.03

The specific energy content of lead accumulators and various fuels in comparison to petrol (n-octane as the model chemical compound). Except for the accumulator, the tank container and the volume of the energy conversion device are not included in the values shown.

Renewable Energy. Edited by R. Wengenmayr, Th. Bührke. Copyright © 2008 WILEY-VCH Verlag GmbH & Co. KGaA, Weinheim. ISBN 978-3-527-40804-7

Capacity of Various Energy Storage Devices" on p. 96). Another advantage of a fuel-cell power train is its more favorable efficiency profile in the partial load regime. For applications in vehicles, fuel cells with lower operating temperatures are especially desirable. They produce sufficient power even at ambient temperatures and thus permit short start/stop cycles while maintaining a long operating life.

Very small fuel cells with power outputs in the range of a few watts are being developed for another attractive application: as power supplies for portable electronic devices such as mobile phones, camcorders or laptop computers. The requirements for efficiency and cost are less critical here than in the other application areas. The combination of a micro fuel cell with a small supply of methanol or hydrogen promises to provide a considerably longer operating time than the lithium batteries currently used.

At present, many new, sometimes spectacular results are being reported from these three areas of applications for fuel cells. We want to shed some light on the facts behind the euphoric news stories and PR strategies, describe the current state of development, and show which problems still remain to be solved.

Fuel Cells for Road Vehicles

For vehicular applications, the demands on a fuel-cell system are very rigorous. Especially critical factors in comparison to stationary applications are the necessary smallest-possible volume and weight for a given power output. Furthermore, a fuel-cell powered drive must be adapted to short operating times, which is the reason that only low-temperature fuel cells are suitable. It must be ready to deliver power within a few seconds and to follow rapid changes in power demand during acceleration and braking of the vehicle. Along with the technical hurdles, the cost factors are very stringent in the vehicular applications area. In order to be competitive, the costs must not be higher than those of the established technology using internal-combustion engines.

One advantage of the "Polymer-Electrolyte-Membrane Fuel Cell" (PEMFC) compared to other low-temperature fuel cells is its straightforward cold-start behavior. At 20° C (68 F), they already deliver about half their maximum power, and they reach their optimal operating temperature in less than a minute. In addition, the polymer membrane which serves as electrolyte is mechanically robust and in general not sensitive to rapid temperature changes. Thus, the PEMFC is predestined for non-continuous operating modes.

The system can furthermore react to power-demand changes in the millisecond range, as long as the supply of fuel is guaranteed. Thanks to its compact construction, sufficiently small volumes and weights per unit of output power can be achieved.

A clear weak point of the PEMFC consists in the fact that the membrane must contain water in the liquid phase. Only then is the membrane proton conducting and the elec-

trochemical reaction can proceed. Drying out of the membrane can be prevented by employing technical "tricks", for example by moistening the reaction gases before their entry into the cell. This limits the starting temperature to above 0° C (32 F) or requires specific measures to prevent freezing. The highest operating temperature, at least at low-pressure operation, is limited to well below 100° C (212 F). This small range of usable temperatures however introduces new problems. One difficulty occurs when the PEMFC is operated with methanol or gasoline instead of pure hydrogen. During reforming of these fuels to give hydrogen and carbon dioxide, carbon monoxide is also produced. Particularly at temperatures below 100 °C (212 F), it poisons the noble-metal catalysts in the anode and thereby hampers the H_2 conversion reaction. In order to prevent the output power from decreasing dramatically, the CO impurities must be removed from the anode gas before it is fed to the cell.

Another problem is the requirement of a more elaborate cooling system compared to an internal-combustion engine. The latter, due to its high operating temperature, can eliminate two-thirds of its waste heat via exhaust gases and thermal radiation, so that only one-third need be carried off by the cooling system.

PEMFC systems have in the meantime reached the stage of prototype construction. Currently, H_2/O_2 systems in the power output range above 300 kW (Figure 2, below) and reformate/air systems of up to 250 kW (Ballard) have been implemented. Reformate refers to the product of the

Fig. 1 *A low-floor bus from the MAN corporation, with an H_2 fuel-cell power train. It was introduced in the year 2000; the fuel-cell system and electrical components are built by Siemens, the H_2 storage system by Linde, and the project design was by Ludwig-Bölkow-Systemtechnik GmbH.*

INTERNET

Basics and history of fuel cells
americanhistory.si.edu/fuelcells/basics.htm

Hydrogen and fuel cells
www1.eere.energy.gov/hydrogenandfuel-cells/fuelcells/index.html

Fundamentals and applications of fuel cells
www.nfcrc.uci.edu/2/Default.aspx

Ballard Power Systems (PEM)
www.ballard.com/

Siemens Stationary Fuel Cells (SOFC)
www.powergeneration.siemens.com/products-solutions-services/products-packages/fuel-cells/

Research Center Juelich (PEMFC, DMFC, SOFC)
www.fz-juelich.de/ief/ief-3/fuel_cells/

H_2 storage
www.eoearth.org/article/Hydrogen_storage
www.ifres.ch/Homepage/DB/Paper4.pdf

FIG. 2 | **THE HISTORY OF FUEL CELLS**

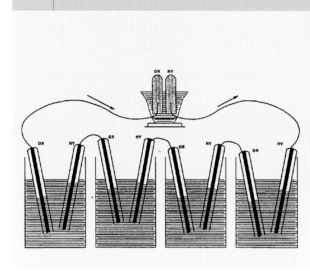 

so-called reforming, i.e. the chemical conversion of hydrocarbons such as methane or natural gas to hydrogen (see also the infobox "The Production and Storage of Hydrogen" on p. 102).

Prototype fuel-cell powered buses (MAN, Neoplan) and automobiles (e.g. Daimler, Opel, Ford, Toyota, etc.) have been tested for several years (Figure 1). In earlier long-term tests of the fuel-cell systems, operating lifetimes of more than 10,000 hours were confirmed.

The performance of current PEMFC prototypes and the experience gained would seem to demonstrate that this technology is in principle mature and requires only adaptation for installation in mass-produced vehicles, if it were not for the high manufacturing costs. They compel great efforts of cost-reduction. In addition, the exotic fuel hydrogen will require not only an elaborate new vehicular technology, but also a completely new infrastructure from its production through the establishment of a network of service stations for fuel delivery. The costs of current PEMFC prototypes are still more than 10,000 US dollars per kilowatt of electric output power. In order to compete with internal-combustion engines, the overall costs must be lowered to around 30 US dollars per kilowatt.

One path towards lower costs involves applying mass-production methods to manufacture the cells. This includes both individual components and also the overall system of the cells and their ancillary devices, which requires considerable simplifications. However, a transition to automated mass production alone will not yield the necessary cost reductions of more than two orders of magnitude. For this, the costs of the materials employed must also be drastically lowered, and this in turn will require considerable progress in materials research.

Less expensive alternatives to the currently-used materials must be developed and tested. Where this is not possible, the amounts of the materials used will have to be reduced as far as possible.

In the case of the PEMFC, for example, optimisation of the electrode structure yielded a reduction of platinum loading from previously 20 g/kW (0.7 oz/kW) to values near 1 g/kW (0.04 oz/kW). Expensive membrane electrolytes can be replaced by cheaper alternative materials, and finally, in future the cell frameworks will be fabricated of common and therefore cheaper construction materials, such as Fe-based alloys or moldable plastics.

A first experimental study has shown that the manufacture of less expensive fuel cells is possible (Figure 3). The next step will be the construction of prototypes, which can be tested under realistic operating conditions for the intended applications.

If it proves to be possible to implement all these ideas without reducing the performance of the cells, then the ambitious goal of a drastic cost reduction can be achieved. While it is to be assumed that the basic technical problems can be solved, establishing a hydrogen supply is a still open question. Before the economic considerations can be dealt with, there are yet many unsolved technical problems. These relate to the demands on automobile power trains (dynamics, start-up behavior, etc.), on the fuel (purity), and on the infrastructure for fuel production and distribution up to the end users at the service stations.

Which Fuel is Most Suitable?

Hydrogen in liquid form – at a temperature of –253 °C (–423 F) – has the highest energy density of any fuel. However, its everyday handling would not be simple. It requires expensive cryogenic tanks which must be protected against accidents and, for security reasons, need automatic filling systems. Other methods, such as storage of hydrogen in pressure vessels at 250 up to 700 bar (0.33-0.93 torr), or in metal hydrides, are also complicated and costly.

Since natural gas, petrol or methanol is relatively simple to transport and to fill and store in tanks, these fuels represent an attractive alternative to hydrogen. By reforming,

Fuel cells then and now: from the initial experiments by Sir William Grove in the year 1839, through the first technical single cells (1965) and on to the world's largest PEM installation (H₂/O₂) at present, dating from the year 1998.

these hydrocarbons can be converted into hydrogen, and the existing distribution infrastructure could be readily adapted to them. Reforming can be carried out either decentralized at the service stations using natural gas (or petrol), or else directly in the vehicle using petrol (or methanol).

Steam reforming is already widely used industrially; there, however, it is carried out under constant, controlled operating conditions. Thus, its system-specific peculiarities,

namely a large volume, poor flexibility in terms of regulating range and dynamics, and long preheating times are no drawbacks. For vehicular use, however, these properties present serious problems. Some of them can be alleviated, for example the long preheating times, by using partial oxidation or autothermal reforming instead of the more efficient steam reforming method. Then, however, a decrease in the hydrogen yield must be accepted.

HOW DO FUEL CELLS WORK?

Fuel cells are electrochemical energy converters. They convert the energy of a chemical reaction directly, i.e. without a thermo-mechanical intermediate step, into electrical energy. Normally, in an (exothermal) chemical reaction, the electric charges (electrons) are exchanged directly between the reacting atoms or molecules. So, for example, hydrogen reacts spontaneously with oxygen in the **detonating gas reaction**; the large amount of energy released by the oxidation of hydrogen is completely converted into thermal energy.

The trick in the fuel cell consists of not allowing the "fuel" to react directly with the atmospheric oxygen, but rather making it first give up electrons at the anode. Via the external circuit, the electrons flow through the power-consuming device and return to the cathode; there, they are taken up by the other reaction partner, typically oxygen from the air. In this way, the reaction can be carried out in a controlled manner and with a high yield of electric current.

The operating principle can also be considered to be a **reversal of electrolysis**. The construction of a fuel cell is very similar to that of a battery: It consists essentially of two electrodes

which can conduct electrons (anode and cathode) that are separated from each other by an electrolyte, which is an ionic conductor. The main **difference from a *battery*** consists of the fact that in the latter, the electrical energy is stored chemically in the electrodes, while in a fuel cell, the energy carrier is stored externally and the electrodes merely fulfil a catalytic function for its reaction. The structure of a fuel cell is shown schematically in the Figure using the example of a polymer-electrolyte-membrane fuel cell. The following chemical reactions take place in it:

Anode reaction:
$$2 H_2 \rightarrow 4 H_{ads}$$
$$4 H_{ads} \rightarrow 4 H^+ + 4 e^-$$

Cathode reaction:
$$O_2 \rightarrow 2 O_{ads}$$
$$2 O_{ads} + 4 H^+ + 4 e^- \rightarrow 2 H_2O$$

Overall reaction:
$$2 H_2 + O \rightarrow 2 H_2O$$

The simplest and preferred reactants are hydrogen as fuel and oxygen as oxidant. Purified

reformer gas (CO content < 100 ppm!) can also be used on the cathode side and air on the anode side. In technical systems, the fuel cells are connected together into "stacks", in order to obtain higher electric voltages and more power. Fuel-cell systems have a high efficiency, which in principle can be well above that of internal-combustion engines.

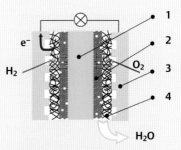

Structure of a PEMFC, schematic.
1: Polymer electrolyte,
2: Pt Catalyst,
3: Cell frame,
4: Current collector and gas diffusion layer

Fig. 3 *An experimental study: A low-cost fuel-cell stack of PEMFCs with an output power of about 10 kW. This project was supported by the European Union (Grant no. JOE3-CT95-0027).*

Additional equipment is also required for cold starting, rapid load variation, and gas purification; thus far, there are no approved designs which are vehicle compatible for these processes. An additional problem is the space requirement for the reformer and the purification stage. Packing them under the hood of an automobile is an ambitious developmental goal. It is even more difficult to guarantee their control over a wide range of loads at simultaneously short response times. To tide over these shortcomings, an electrical storage device such as a high-power battery or a capacitor could be operated in parallel with the fuel cells. A complete power train is shown schematically in Figure 4.

The advantages offered by fuels that are liquid at room temperature are obtained only at the price of introducing a number of new difficulties. Some of these could turn out to be deadly for the "reformer route" in later large-scale applications. A variant on the PEMFC might be the direct methanol fuel cell. It allows direct reaction of methyl alcohol without previous treatment in a reformer or subsequent carbon monoxide purification. However, this development

FIG. 4 | A FUEL CELL POWER TRAIN

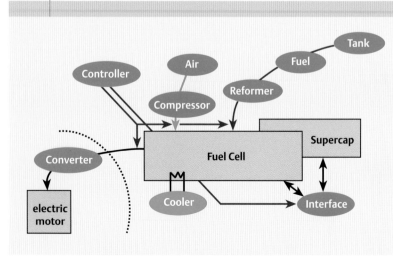

System concept of a PEM fuel-cell power train for road vehicles.

is still in the laboratory stage, and here, again, a drastic cost reduction and improvement in performance will be required.

Fuel Cells for Stationary, Decentralized Power Supplies

The requirements on fuel cells for stationary power supplies can be more easily fulfilled. In this case, we are not considering large central plants with an output of several 100 MW, since these can be more cost-effectively constructed using conventional technology, and they also operate with a high electrical efficiency (up to nearly 60 %). Fuel cells offer advantages in the power range below 100 MW, which is typical of decentralized power generation. In this range, the investment costs for conventional plants relative to their output power increase rapidly, and at the same time their efficiency decreases. Here, system efficiencies of more than 40 % in the 100 kW class and of up to 70 % for plants in the range of 10 MW are possible (using natural gas as fuel).

Such high efficiencies can be obtained when the waste heat from the fuel cells is available at a temperature above 500° C (932 F), so that a subsequent gas and steam turbine cycle can deliver additional power. If the operating temperature is lower or if the plant is so small that such expensive peripheral equipment would not pay, then steam is produced as a by-product for industrial processes or district heating networks. In small systems (<100 kW) the fuel cell supplies in addition to electric power only warm water for local heating or cooling.

Operating temperatures above 600° C (1,112 F), which are typical for the so-called high-temperature fuel cells, first of all have the advantage of fast electrode kinetics so that expensive noble-metal catalysts can be omitted and the cells can tolerate carbon monoxide. Therefore, in principle, gasified coal or biomass can be used as fuels. Natural gas with a high methane content is particularly attractive. This fuel can be reformed within the fuel cell stack itself, making use of the heat produced in the cells. Then an externally heated reformer is not required, which saves on investment costs and fuel, and furthermore the power requirements for the cooling cycle are lower. This leads to an improvement in efficiency in comparison to the PEMFC and PAFC systems, which operate at lower temperatures. On the other hand, higher operating temperatures have the disadvantage that the materials of the cells and the system components must be more corrosion-resistant and mechanically stable, which makes them more expensive.

In contrast to automobile power trains, where by general agreement the required specifications can be met by only one type of cell, the PEMFC, for stationary applications four different systems are currently being tested: the PEMFC, the PAFC, the MCFC, and the SOFC.

The PEMFC

Decentralized applications of the PEMFC for combined heat and power (CHP) production are limited by the low level of waste heat from this cell type. Since its waste heat is sufficient for low-temperature heating systems, PEMFCs could be utilized as household energy suppliers.

Although PEMFC systems require a costly fuel pretreatment which reduces their overall efficiency, their suitability for compensating peak load power requirements, or as household energy supplies, is currently being investigated. In Germany in June 2000, a first field trial was begun by the energy company BEWAG (Vattenfall) in Berlin: A 250 kW$_{el}$ installation built by Alstom/Ballard supplies the electric power needs of a heating plant. Limited by the energy consumption of the natural-gas reformer, the system exhibits an overall efficiency of 35 %. This is not superior to conventional generators driven by internal-combustion engines, and the investment costs are still an order of magnitude too high. For energy supplies to individual dwellings, small installations with 5 kW of electric and 7 kW of thermal output power are under test. Here, again, the ability to compete with established technology is not yet in view.

The Phosphoric Acid Fuel Cell (PAFC)

Owing to their higher operating temperature of 200° C (392 F), phosphoric-acid fuel cells can tolerate a considerably higher CO concentration of 1–2 % in the fuel gas than the PEMFC. This means that an elaborate purification of the fuel gas downstream from the natural-gas reformer is unnecessary. Of all the types of fuel cells which are candidates for stationary energy generation, the PAFC is the most technically mature.

Thus, ONSI (USA) has already put a complete system (including reformer) with a power output of 200 kW on the market, and it has proved itself over years of operation. Unfortunately, here again the electrical efficiency reaches not more than 40 %, and thermal energy is output only in the temperature range around 90° C (194 F). At 5,000 US dollars per kW$_{el}$, the price is still far above that of conventional technologies.

TYPES OF FUEL CELLS

Fuel cells can be classified as low-temperature and high-temperature cells. Low-temperature fuel cells operate in a range from room temperature up to about 120° C (250 F), while high-temperature fuel cells require an operating temperature between 600 and 1,000° C (1,100-1,800 F).

The phosphoric-acid fuel cell, with an operating temperature around 200° C (390 F), lies between these two rough ranges. Fuel cells are named for the type of electrolyte they use. There are alkaline fuel cells (AFC), the PEMFC (polymer-electrolyte membrane fuel cell), the molten carbonate fuel cell (MCFC), the phosphoric acid fuel cell (PAFC), and solid oxide fuel cell (SOFC). Only the direct methanol fuel cell (DMFC) does not mention the electrolyte in its name, but instead indicates its ability to react methanol directly as fuel. Typically, it is based on a PEMFC. The different electrolytes determine to

a large extent the characteristic properties of the corresponding fuel cells, such as their operating temperatures and conductivity mechanisms. Directly related to this is the choice of usable catalysts, requirements for the process gas, etc.

Low-temperature fuel cells require noble-metal catalysts in their electrodes (platinum or noble-metal alloys), in order to activate the electrochemical reaction. These catalysts are in general sensitive to CO poisoning at low temperatures. The higher operating temperature of the PAFC, in contrast, allows it to tolerate CO concentrations of 1–2 %, so that instead of pure hydrogen, so-called reformer gas can be directly feed to the cells.

The high operating temperatures of the MCFC and the SOFC make these cells insensitive towards CO and even allow the direct reaction of methane (natural gas).

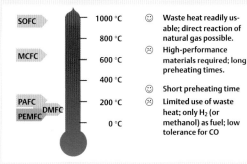

Typical operating temperatures of the different types of fuel cells and the properties which follow from them.

FIG. 5 | SOFC TUBE CELLS

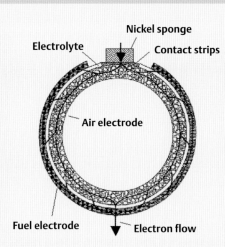

Left: Block of 12 bundles of SOFC tube cells (Siemens-Westinghouse); Right: Cross section through a single cell.

Using methane as an example, we show the various methods for so-called reforming, whereby hydrogen is produced from hydrocarbons. Industrially, steam reforming is applied technically on a large scale.

It employs an endothermal reaction with water vapor, for which heat must be input. The resulting gas mixture is called reformer gas. Its main components are hydrogen, carbon dioxide, water vapor and about one percent of carbon monoxide. Partial oxidation (POX) requires no heat input. In this case, the methane is oxidized substoichiometrically using suitable catalysts.

The energy balance sum of these two variants is termed autothermal reforming. Operational differences, besides the time for preheating, lie in the different yields of H_2. As byproduct of the reaction, CO is formed. The CO content at the output of the reformer is lowered by subsequent "shift" and purification stages from initially over one percent to values below 100 ppm (parts per million). Methanol or long-chain hydrocarbons (petroleum) can also be reformed. However, in the case of petroleum reforming, the H_2 yield is reduced owing to the less favorable C/H ratio to about 38 % (autothermal reforming) or even to only 30 % (POX).

For H_2 storage, there are established methods using pressure vessels or cryogenic tanks; metal hydrides could offer an alternative. They permit the reservoir to operate almost without pressure; the maximum H_2 capacity relative to the mass of the storage medium is, however, only 1.5 to 1.8 %. Hydrides with better storage behavior (max. 6 wt.-%) are known, but they release the absorbed gas only above about 200° C.

Steam Reforming:

$$CH_4 + 2\,H_2O \rightleftharpoons CO_2 + 4\,H_2$$

Partial Oxidation:

$$CH_4 + O_2 \rightleftharpoons CO_2 + 2\,H_2$$

Autothermal Reforming:

$$CH_4 + x\,H_2O + (1-x/2)\,O_2 \rightleftharpoons CO_2 + (2+x)\,H_2$$

Various methods of reforming methane. The reaction schemes are idealised.

TAB. | REFORMING METHOD

	Steam Ref.	Autothermal Ref.	Partial Oxid.
ΔH / kJ mol^{-1} at 25 °C (77F)	165	0	– 318
Reformate / % H_2	74	53	41

Comparison of possible reforming routes for methane: Energy balance and typical hydrogen concentration.

The Molten Carbonate Fuel Cell (MCFC)

The MCFC takes its name from its electrolyte of alkali carbonates, which is a melt at the operating temperature of 650° C (1,200 F). The current is carried in the electrolyte by CO_3^{2-} ions. The peripheral system components such as connecting tubes and heat exchangers require only a limited high-temperature serviceability; however, the extremely aggressive electrolyte causes corrosion of the cell components. Maintaining the CO_3^{2-} charge transport within the electrolyte requires CO_2 circulation between the fuel gas exhaust and the air feed.

MCFC installations with output powers of up to 2 MW have already been built, and efficiencies between 40 and 50 % were measured. Here, again, the goals for operating lifetimes and costs have not yet been met.

The Solid-Oxide Fuel Cell (SOFC)

In the SOFC, a thin ceramic layer of yttrium-oxide-doped zirconium oxide (YSZ) serves as electrolyte, and it allows the passage of oxygen anions at an operating temperature of 900 to 1,000° C (1,650-1,830 F). Thus far, various planar and also cylindrical cell concepts have been developed. Figure 5 shows on the right the basic structure of a cylindrical ceramic single cell in cross section. The inside of the tube, which is 1.5 m long (5 ft) and 2.2 cm thick (0.9 in), contains flowing air, while the fuel gas passes along the outer wall. The cell can deliver 100 to 200 W. In the SOFC generator shown on the left side of Figure 5, 1,152 such tubes are connected together.

The power output of the systems currently being tested ranges from one kilowatt for household power supply (electrical efficiency: 30 %) up to a pressurized 200-kW unit with a gas turbine and 53 % efficiency. So far, a 100-kW plant has been operated for three years without showing any ageing.

With an advanced SOFC plant concept, it is hoped that electric power can be generated from natural gas without carbon dioxide emissions into the atmosphere. Up to now, for technical reasons only 85 % of the fuel gas is burned in the cells. The remaining gas is burned with air before leaving the system. Therefore, the final exhaust consists of a mixture of CO_2, H_2O and N_2. If the fuel utilization is increased to 100 %, then after condensing the water vapor, the exhaust will consist of pure CO_2. It can be pumped into explored natural gas repositories and thus no longer burdens the atmosphere.

Marketing of the SOFC technology is to be expected only in the medium term, in spite of the performance already achieved and the attractive perspectives for the future, since its manufacturing costs are currently still two orders of magnitude higher than the goal for market entry. Thus, the most important task in the coming years is here also cost reduction.

Micro Fuel Cells

For supplying low to very low power levels (1–100 W), fuel cells are also interesting, because they can make the high energy densities of hydrogen or methanol available to electric consumers. If very long operating times are desired, then the combination of a fuel storage system and a fuel cell has advantages compared to a battery. Operation at room temperature and their rapid operational readiness would seem to recommend the PEMFC as the only reasonable solution for these applications.

Initial demonstration experiments have been successful, and they promise practically a doubling of the operating time of a laptop, for example. Nevertheless, from the current standpoint it will be difficult for PEM micro fuel cells to displace the established battery technology. In particular, the question of recharging, and the logistics of the hydrogen cartridges are still unsolved. Only a considerable increase in operating times (by a factor of 10) in comparison to storage batteries could compensate for this drawback. A possible alternative could be offered by metal hydride storage systems for hydrogen, but the currently available technology is not yet satisfactory. The solution could be the direct-methanol fuel cell (DMFC). It can operate directly with methanol as fuel instead of H_2; methanol is liquid at room temperature and atmospheric pressure and can thus be easily stored. There are however still several technical hurdles to be overcome before the DMFC is ready for applications. Among other things, self-discharging in the switched-off state must be reduced, and the long-term stability of the electrodes must be improved.

If there is a breakthrough in hydrogen storage or a quantum leap in the technical maturity and concept simplification of the DMFC, then a market for micro fuel cells could be developed much more quickly than for vehicle power trains or for decentralized energy supplies.

Outlook

Whether, and how soon, fuel-cell vehicles will take over the streets depends to a large extent – along with cost reductions – on how rapidly the problems of hydrogen supply can be solved and can be adapted to the technical requirements of vehicular power trains. Applications in stationary power generation, where dynamics, cold starting and H_2 purity are less important, will depend mainly on the achievable cost decreases. Or will new H_2 storage systems appear and help the micro fuel cell to become a high flyer? Prognoses are difficult here.

There will be a suspenseful race among these three fundamentally different applications of fuel-cell technology to conquer the markets.

Summary

Fuel cells have reached a high level of technical development. The PEMFC has demonstrated its reliability for a series of niche applications, as well as in the form of prototypes for mobile and decentralized use. The SOFC and the MCFC have already been tested in plants of 100 kW and more. However, in order for fuel cells to be economically competitive with the established technologies for mobile and decentralized energy conversion, a drastic cost reduction must be achieved, both for the fuel-cell stack itself and for the ancillary systems required for its operation. In vehicle applications, the still open questions of fuel supply (infrastructure, H_2 production and H_2 storage) must also be addressed.

References

[1] L. Carette, K.A. Friedrich, U. Stimming,
Fuel Cells **2001**, *1*, *5*.
[2] Fuel Cell Systems Explained
J. Larminie, A. Dicks, Wiley 2000.
[3] Handbook of Fuel Cells (4 volumes)
W. Vielstich, H.A. Gasteiger, A. Lamm, Wiley 2003.

About the Authors

Manfred Waidhas, born in 1958 in Nuremberg, studied chemistry in Erlangen and obtained his doctorate in physical chemistry there in 1985; since 1985 he has been with Siemens, specializing in electrochemical energy conversion and storage, in particular low-temperature fuel cells.

Harald Landes, born in 1954 in Nuremberg, studied physics in Erlangen and completed his doctoral studies in physical chemistry there in 1985; he has been at Siemens since 1985, specializing in electrochemical energy conversion and storage, in particular high-temperature fuel cells.

Contact:
*Dr. Manfred Waidhas, Dr. Harald Landes,
Siemens AG, Corporate Technology,
Department of Electrochemistry,
Günther-Scharowsky-Str. 1, D-91058 Erlangen,
Germany.
manfred.waidhas@siemens.com
harald.landes@siemens.com*

Solar Air Conditioning

Cooling with the Heat of the Sun

BY ROLAND WENGENMAYR

When the Sun burns down mercilessly, it heats up buildings especially fast. But precisely then, it also delivers the most energy for powering large air-conditioning systems.

In order to keep cool, the Americans in particular consume enormous amounts of energy. In 2001, the air-conditioning systems of the roughly one hundred million households in the USA ate up nearly five percent of the annual US electrical energy production. Worldwide, the market for air conditioning is growing at a startling pace. Along with it, the consumption of environmentally harmful fossil fuels is exploding. But there is a way out: solar energy can also be used for cooling.

When the Sun is shining mercilessly, it not only makes people suffer from the heat, it also provides a large amount of technically usable thermal energy. What would be more obvious than to use this energy directly to power the air conditioning? Such installations would offer an ideal solution for households and industrial buildings, "whose cooling requirements are determined mainly by the climate, that is the momentary solar irradiation and the air temperature ", says Volker Clauss of SK SonnenKlima GmbH in Berlin. They are also interesting for houses which have no connection to the power grid. This is true for example in Australia, since there, only ten percent of the land area is served by the electric power network. A high percentage of the households supply their needs by using a diesel generator. In these houses, cooling by utilizing solar energy would yield clear-cut reductions in diesel fuel consumption and CO_2 emissions.

Even in regions with a good infrastructure, an alternative to conventional air conditioning can pay off. In China, the Arabian States or around the Mediterranean, for example, there is a large market potential – and even in cooler Central Europe. "At present, in Europe about 120 systems are installed, with a total cooling power of 12 megawatts, more than a third of them in Germany", reports Hans-Martin Henning, scienctist of the Fraunhofer Institute for Solar Energy Systems (ISE) in Freiburg, Germany. The ISE researchers for example constructed a solar-assisted air conditioning system for the University Clinic in Freiburg in 1999; it cools a laboratory building.

All solar-powered air conditioning systems operate on the same basic principle: an evaporating liquid takes up heat and cools its surroundings; if the vapor is pumped away and liquefied elsewhere, it gives up its stored thermal energy again. Using these two steps, heat can be transported out of a closed room. Refrigerators and conventional air conditioners also operate on this principle. In their case, a strong electric compressor densifies the vapor of the cooling medium, so that it becomes liquid and frees its thermal energy outside the space to be cooled. A solar-powered air conditioner, in contrast, has to get along without an electric compressor – after all, solar collectors do not generate electric power; instead, they yield only heat. These systems cannot operate entirely without electric pumps to move the cooling medium, but the pumps can be designed to be much less powerful than a compressor. Thus, even a relatively small photovoltaic installation can provide the necessary electric power.

The function of the compressor is taken over by so-called absorbers: these bind the vapor of the cooling medium – it is usually water – and give off the heat content of the vapor. When heated, they release the water again and thereby cool their surroundings. There are a number of solid and liquid absorber materials; in everyday life, we often see them in the form of silica-gel beads, which keep moisture-sensitive products dry in their packages.

Closed cooling cycles

In order to use the absorption effect for solar-assisted air conditioning, researchers and engineers have developed several technical concepts. Experts class them as systems with open or with closed cooling cycles. Systems with a closed cycle pump cold water through cooling tubes in the ceiling and walls of a building, like a reverse heating system. For systems of this type, SK SonnenKlima has developed a small refrigeration plant which, with a maximum of 16 kilowatts of cooling power, can cool for example one floor of a hotel.

This absorption refrigeration plant feeds a separate cooling-water circuit via a heat exchanger. To produce the cooling power, it evaporates water in a chamber at low pressure and five degrees Celsius (41 F). The absorber medium, a concentrated solution of the salt lithium bromide, then absorbs the water vapor, after which it is pumped into a second chamber at a higher pressure. There, the actual "motor" of the cooling system, namely the heat from the solar collector, again separates the water from the salt solution. A portion of the cooling power has to be diverted to condense the water vapor released from the salt, after which it passes again into the low-pressure chamber, completing the cy-

Renewable Energy. Edited by R. Wengenmayr, Th. Bührke. Copyright © 2008 WILEY-VCH Verlag GmbH & Co. KGaA, Weinheim. ISBN 978-3-527-40804-7

FIG. 1 | SOLAR COOLING WITH A DOUBLE BOTTOM

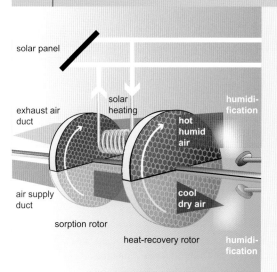

An open "absorption-assisted cooling plant" has an air supply duct (below, blue color) and an exhaust air duct (above, red). Two rotors turn very slowly through both ducts, the so-called sorption rotor and the heat-recovery rotor. Both have a honeycomb structure with many fine channels, in order to present the largest possible surface area to the air which is flowing past. The aborption rotor initially dehumidifies the incoming air. This dry air can now take up a large amount of water from the humidifier which follows. Since the water must evaporate in this process, it removes heat from the air: it cools and can be used for the air conditioning of the building. In order that the sorption rotor be continuously ready to take on water, the used exhaust air from the building is heated by a solar collector. Then it can take on additional moisture in flowing through the sorption rotor and thus carry off the water absorbed from the incoming air. The heat-recovery rotor and a second humidifier increase the cooling power of the plant.

(Graphics: © Roland Wengenmayr)

cle. Closed-cycle refrigerators of this type require only moderate solar actuating temperatures between 55 and 100 degrees Celsius (130-210 F). "So they can be operated with low-cost flat solar panel collectors", Clauss emphasizes.

The Berlin firm aims at a market price of under 10,000 Euro (approx. 14,400 $) for their cooling plants from 2010 on; added to that is the investment for the solar collectors: all together, a small plant of this type is still more expensive than an electric compression cooling system of the same cooling power, which costs between 4,000 and 6,000 Euro (approx. 5,800-8,600 $). However, it requires only one-tenth as much electric power, and is therefore much cheaper to operate.

Georg Buchholz, general manager of SK SonnenKlima, envisions a strong future market for small closed-cycle systems. In his view, in addition to North America, Asia and the Mediterranean region, also Arabian countries such as the United Emirates will be interesting markets for sales of the plants – but also Central Europe. "We will have no problems in selling our planned production series", says Buchholz optimistically.

Open cooling cycles

Open systems draw in air, and not only cool the gas mixture, but also dehumidify it at the same time (see Figure 1). They are suitable only for large central air-conditioning systems with a high air throughput which require several hundred kilowatts of cooling power. Instead of using solar energy, plants of this type can make use of industrial waste heat or other low-temperature heat sources for their actuation. Thus, components for open systems are already on the market. For example, the factory of the furniture castor producer H. C. Maier GmbH in Althengstett, on the northern edge of the Black Forest, has been air conditioned since 2000 using a solar-assisted open-cycle cooling plant.

The researchers at the Fraunhofer institute, together with industry partners, installed a solar-assisted air conditioning system for several rooms of the Chamber of Commerce and Industry of South Baden in Freiburg in 2001. This plant uses an open absorption technique. The technology is still in its fledgling stage, says Hans-Martin Henning of the ISE institute, but it will mature rapidly and will capture the market.

Summary

When the Sun is burning down, it heats up buildings very fast. But at the same time it delivers the most energy, which can be used to power large air-conditioning systems. At present, there are two basic methods of operating solar-assisted cooling plants. Closed systems work with a closed-cycle cooling-water circuit. They are principally suited for single floors or smaller buildings. For large buildings with a high throughput of air, open systems are more efficient: They cool and dehumidify the air supply directly. Both methods utilize absorption media – i.e. drying agents – which remove water from the air and thereby also decrease its thermal energy. Afterwards, the absorption media are again freed of the water they have taken up by applying solar heat. This principle permits heat to be transported out of the buildings.

About the Author

Roland Wengenmayr is editor of the German physics journal "Physik in unserer Zeit" and a science journalist.

Contact:
Roland Wengenmayr,
Physik in unserer Zeit,
Konrad-Glatt-Str. 17,
D-65929 Frankfurt am Main, Germany.
Roland@roland-wengenmayr.de

Climate Engineering

A Super Climate in the Greenhouse

BY ROLAND WENGENMAYR

Fig. 1 In the new Post Office Tower in Bonn, Germany, the engineers for the first time implemented a decentral air-conditioning plant in a high-rise building. Between the two semicircular segments which are offset from one another, there are "sky gardens" that extend over several stories and fulfill important functions for heat transport. (Photo: Deutsche Post World Net.)

Not only glider pilots, but also climate technologists know how to get the most out of winds and updrafts.

Modern large buildings often present impressive glass facades; but without air conditioning, their interiors would be unbearably hot on sunny Summer days, and would be too cold in Winter. The heat output of human beings is also a factor which is not to be underestimated when many people occupy a building; furthermore, one must not forget the energy output from technical apparatus such as computers. In conventional high-rise buildings, air conditioning equipment thus takes up about every 20th floor, and on all the other floors there are voluminous air ducts above the false ceilings. This air conditioning eats up money and energy, and it causes health problems for many of the occupants of the buildings.

A modern architectural concept, in contrast, should make expert use of sources of heating and cooling from the environment, as well as various physical effects which can contribute to providing a basic air conditioning. Additional equipment can then carry out the 'fine tuning' and compensate for peak loads.

This equipment can be much smaller and more energy economical. Such an alternative construction represents to be sure an enormous technical challenge, since it must provide comfortable conditions in hundreds of rooms reliably during every season of the year. For this reason, architects work closely together with air conditioning and climate experts.

The firm Transsolar Energietechnik GmbH, with its central office in Stuttgart (Germany) and a subsidiary office Transsolar Inc. in New York, is such a partner. Its engineers and physicists belong among the pioneers of a sustainable air-conditioning technology. The mechanical engineer Matthias Schuler and some of his research colleagues founded the young venture in 1992, coming from the University of Stuttgart – at that time with the goal of bringing solar energy indoors. Solar collectors play only a minor role today at Transsolar, but the Sun itself is still an important factor in their air-conditioning concepts. After over ten years, Transsolar can exhibit an impressive list of reference projects. It runs from the Mercedes Museum in Stuttgart to the new international Suvarnabhumi airport in Bangkok, whose air-conditioned halls and passageways have to deal with 30 million visitors every year. Star architects such as Frank O. Gehry or Helmut Jahn regularly cooperate with Transsolar, and Schuler is in the meantime also Lecturer in Architecture at Harvard University.

"When we are on board, all the energetic aspects are considered from the first design draft on", explains Thomas Lechner, one of the firm's partners and Professor for Construction Physics at the Technical College in Kaiserslautern. At the same time, every large building is unique with its own operating mode and aesthetics. Glass facade, roof, atria and stairwells, offices, conference rooms, cafeterias and the basement all become elements of an architectural air-conditioning system. Its task is to maintain the air in the building at a comfortable temperature and humidity and circulate it without creating unpleasant drafts.

Feel-Good Temperature vs. Energy Consumption

Beginning with the earliest planning phase, the engineers from Stuttgart employ elaborate computer models. Their software simulates the interior climate during the day and night, for every weather condition and in all seasons. The properties of the windows, walls and ceilings and even the behavior of the people in the building are taken into account, insofar as they influence the interior climate. For example, the less effectively the architects make use of daylight, the more investment and power must be expended for illumination. Artificial lighting can with unsuitable planning become an important heat source within a building.

The results of the model calculations are graphs which show how the air will flow through the rooms and how its temperature changes during the flow. The engineers can also observe what temperatures the ceilings and walls will take on, an important factor in determining the "subjective temperature" and thus for the comfort of the building's occupants.

In constructing very large buildings, the engineers are often breaking new technological ground, and computer simulations alone are not able to describe the situation correctly. In such cases, Transsolar builds real models of the planned building or its critical sections and tests them under varying weather conditions. If necessary, the engineers even construct a 1:1 mockup of a complete office with a section of the glass facade and let it be "occupied" for several months by monitoring equipment.

The new Post Office Tower in Bonn (Germany) shows what modern climate engineering is capable of (Figure 1). The architects Murphy and Jahn in Chicago designed this new administration building for the German Post Office, and the Stuttgart firm was able to pull out all the stops, working on a 162.5 meter (533 ft) high building with 41 stories for the first time. The result is dryly summarized by Lechner: "It is the first high-rise building with decentral ventilation".

FIG. 2 | AIR CONDITIONING

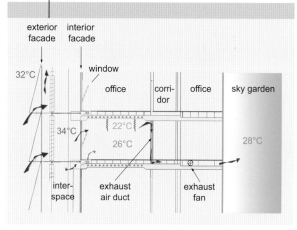

This cross-section through one floor of the Post Office Tower shows the path taken by the fresh air supply through the double facade into the offices (red arrows). From there, it flows as exhaust air under the corridors into the sky gardens, which transport it out of the tower using the chimney effect. Water pipes in the concrete ceilings provide additional cooling for the rooms in Summer (blue arrows); in Winter, hot water is pumped through them for space heating.
(Graphics: Transsolar.)

Instead of a central air-conditioning plant with its enormous air supply and exhaust shafts, the engineers made use of the wind that almost always blows around such a high building that stands alone, and of the chimney effect, which causes warm air to rise in the interior of buildings (Figure 2). In addition, they designed "activated" concrete ceilings: these contain thin water pipes, which carry cooling ground water from two wells below the building in Summer, and hot water for heating in Winter.

The architectural design accommodated this ventilation concept from the outset. The floor plan of the building corresponds to two slightly shifted circular segments, connected by a transitional area. Above the entrance foyer, this area houses five so-called sky gardens; four of them extend over nine stories each, while the fifth and uppermost includes two stories and is reserved for the executive board of the Post. These sky gardens make up the chimney in which the warm exhaust air from the offices rises up and then leaves the tower on the sides.

Wind pushes fresh air into the building through some thousands of openings. In order to be able to use this pumping effect even under extreme weather conditions, the exterior facade consists of a double construction; the air first flows through ventilation shutters into the inter-space between the facades (Figure 3). These shutters are opened or closed depending on the wind velocity, the wind direction and the temperature, and their opening angles are regulated accordingly.

Cooling as in the Orient

In contrast to most high-rise buildings, the rooms of the Post Office Tower have windows in the interior facade

Fig. 3 *The fresh air supply enters through regulated shutters initially into the interspace between the glass facades.* (Photo: Anja Thierfelder.)

which can be opened in order to adjust the interior climate individually (Figure 2). Otherwise, the air flows through the "subcorridor convectors" into the tower walls. These convectors are themselves small air conditioning systems. Ducts conduct the air into the neighboring corridor, where it passes through ventilation slits and finally into a sky garden.

Normally, the wind pressure and the chimney effect are sufficient to provide good ventilation within the whole building. On an average of thirty days each year, a lull in the wind combined with low temperature differences between the inside and outside of the building make additional ventilation necessary, which is provided by fans in the exhaust-air ducts (Figure 2). Of course, it was also necessary to plan for the opposite situation: If a pressure difference between the windward and the lee side were allowed to punch through into the interior of the building during an Autumn storm, then many of the office doors would be pressed shut and desks would be swept clear of papers. But instead, these strong airflows lose their power in the interspace between the two facades, damped by the inlet shutters and the ventilation slits.

In spite of this elaborate design, the contractors were able to construct the building at lower cost than with a conventional air conditioning system, since they gained about 15 percent in the interior volume of the building. To be sure, a portion of the cost savings went into the control systems. Since the middle of 2003, the Post Office Tower has been occupied every day by up to 2000 persons, and the interior climate control operates reliably. "In traditional Arabian houses, the chimney effect has been used for millennia to cool the interior from sixty degrees Celsius (140 F) down to forty degrees (100 F)", explains Thomas Lechner. "We are simply attempting to use the most modern means to achieve twenty-five degrees Celsius (77 F)".

Summary

Large modern buildings with glass facades need extensive air conditioning in summer and heating in winter. In high-rise buildings the air conditioning equipment therefore occupies typically about every 20th floor completely and additionally requires voluminous air ducts above false ceilings on all other floors. Strong central air conditioning causes health problems for many of the occupants of such buildings. An intelligent architecture can overcome this waste of volume, energy and unhealthy environment, as the Post Office Tower in Bonn (Germany) with its 41 stories demonstrates. Its air conditioning uses two basic effects: the chimney effect and the pressure of the wind outside of the 162.5 meter (533 ft) high building. A double facade protects the interior of the building from the strong winds that can blow around a tower of such height. In addition, small decentral air conditioning systems in every office allow its occupants to control their individual feel-good temperature locally. Its concrete ceilings contain thin water pipes which carry cooling ground water from two wells below the building in summer, and hot water for heating in winter. The intelligent climate concept was developed by Transsolar, a German firm that co-operates internationally with renowned architects.

About the Author

Roland Wengenmayr is editor of the German physics journal "Physik in unserer Zeit" and a science journalist.

Contact:
*Roland Wengenmayr,
Physik in unserer Zeit,
Konrad-Glatt-Str. 17,
D-65929 Frankfurt am Main, Germany.
Roland@roland-wengenmayr.de*

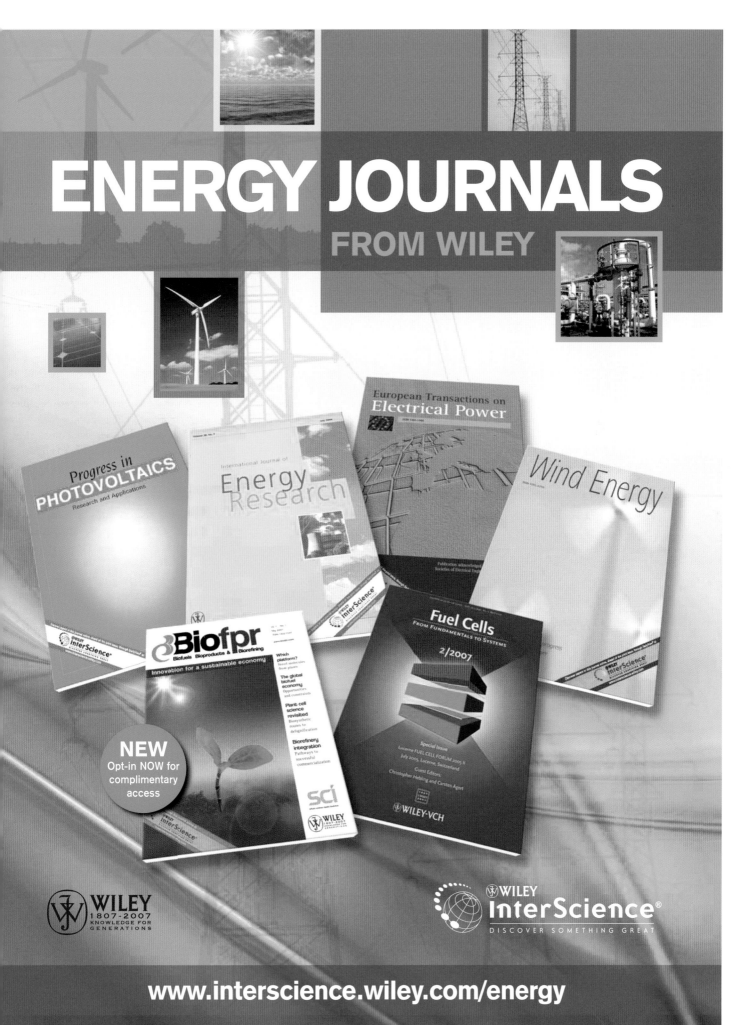

A Low-energy Residence with Biogas Heating

An Exceptional Sustainability Concept

Christian Matt | Matthias Schuler

This example of a therapy center including residential accommodation shows how an intelligent overall concept is capable of cleverly connecting ecology and economics. Furthermore, the combined biofuel and cogeneration unit of the neighboring farmer makes even heating largely autonomous.

The architectural office of Michel, Wolf und Partner (Stuttgart, Germany) developed a complex of buildings in cooperation with our team at Transsolar Energietechnik GmbH (Stuttgart) featuring an exceptional energy concept. It emerged from an old mansion in the small village Bergheim in the north-west of Stuttgart. The Plan was to renovate this villa so it would accommodate a therapeutic institution for a group of just under 40 handicapped people. For accommodating the residents, a new apartment building was attached to the villa, built by Deaconry Stetten. This religious institution wanted a building that would create low energy costs, use regenerative energy, and still provide high living standards for its residents.

The new extension is a long, three-story low-rise building whose generous windows provide an impression of transparency. These large windows fulfill two important tasks in our energy concept. On the one hand, they provide high-quality daylight for all rooms of the new building, thus saving on artificial light and electrical energy. On the other hand, lots of sunlight enters the rooms in winter. This so-lar benefit additionally lowers energy consumption because heating can be reduced.

The fixed budget for the large low-energy home would only provide a feasible solution if we would follow a holistic approach. From the start, all disciplines were included, i.e., structural work, electrical planning, as well as heating, ventilation, and sanitation. For example, the new building rests on a concrete channel instead of individual posts. Half of the channel serves as earth duct, while the other half provides technical supplies such as waste water, water for domestic use, and ventilation.

The earth duct beneath the building supplies the building with fresh air. In winter, it preheats the air naturally before it is brought into the building via a ventilation system. The ventilation equipment provides a sufficient fresh air supply to the residents at all times. At the same time, it is equipped with an efficient heat recovery system that greatly reduces the energy consumption compared to a window-ventilated building (Figure 1). Due to very effective heat insulation and high-quality double-glazed windows in both the new building as well as in the old villa, the buildings lose very little heat in winter, and thus, their demand for heating energy drops.

In summer, thermal insulation and the outside sunshade keep the building cool. However, residents are required to play an active role. Our comfort concept asks them to open the windows themselves during nighttime in order to cool the massive concrete ceilings and walls through this night air flushing. In addition, the earth duct permits that the ground to precool the fresh air before it enters the building on hot summer days.

Heat Supply

Another distinctive feature is the building's heat supply: it has no boiler. Instead, it is connected to the neighboring farm via a local heating duct. The farmer living there has installed a biogas system with a cogeneration unit. He runs an agriculture business with approximately 100 cows whose liquid manure delivers part of the 'fuel'. The remainder comes from organic waste, particularly high-energy fat: this is because this farmer is licensed to dispose of used fat from waste-water oil separators in nearby restaurants and canteen kitchens. From this, the biogas system produces methane gas (Figure 2). The downstream cogen-

FIG. 1 | VENTILATION CONCEPT

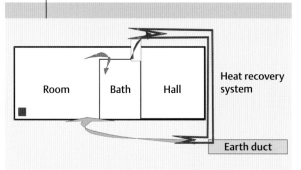

The cross section of the building shows the concept of ventilation via the earth duct and the heat recovery system.

Renewable Energy. Edited by R. Wengenmayr, Th. Bührke. Copyright © 2008 WILEY-VCH Verlag GmbH & Co. KGaA, Weinheim. ISBN 978-3-527-40804-7

eration unit supplies renewable electricity and heat from this methane gas. The farmer sells the electrical energy, amounting to a notable 250,000 kilowatt hours per year, as 'green' energy to his electric supply company according to the fixed price for electricity fed into the grid.

However, the farmer was initially afraid of the investment costs, amounting to a good quarter of a million euros (about a third of a million dollars) that would be necessary for such a system, in spite of his high personal contribution to the construction. Nevertheless, the Deaconry Stetten was able to convince him, since they as operators of the therapeutic institution guaranteed that they would buy from him the heat they needed on a long-term basis. This provides a reliable source of income to the farmer for amortization of his investment costs. It is supplemented with two additional guaranteed sources of income, thus making the project economically feasible.

The farmer can feed the electricity from the cogeneration unit into the grid and sell it; and since this electricity is 'renewable', it is additionally subsidized in Germany via the so-called Act on Granting Priority to Renewable Energy Sources (Renewable Energy Sources Act, see also pp. 113 in this book).

The biogas system features two additional secondary benefits. First, it refines liquid manure to high-grade fertil-izer that the farmer can distribute on his fields throughout the year. Second, the utilization of methane gas in the biogas system prevents this gas, generally produced by cows, from escaping into the atmosphere as is usually the case in agriculture. Methane is a particularly dangerous greenhouse gas, 20 to 30 times more potent than carbon dioxide.

However, the cogeneration unit, particularly during the summer, produces more heat than the Deaconry building needs for heating and hot water, even when the farmer also uses the heat to cover his private heat consumption. This is due to the high heat standards in the new residence and the renovated villa.

Therefore, the farmer is looking for additional customers that could use the heat locally. In order to be able to guarantee the delivery of heat on winter days and if the cogeneration unit breaks down, the boiler in the farmhouse was supplemented with an oil reserve so that it can serve as a backup system.

CO₂ Balance

Apart from investment and operating costs, the CO_2 balance of the possible solutions was very important for finding the best concept. After lowering the energy consumption for heating and hot water by means of the low-energy concept, the heat required for the therapy center amount-

FIG. 2 | HEAT SUPPLY

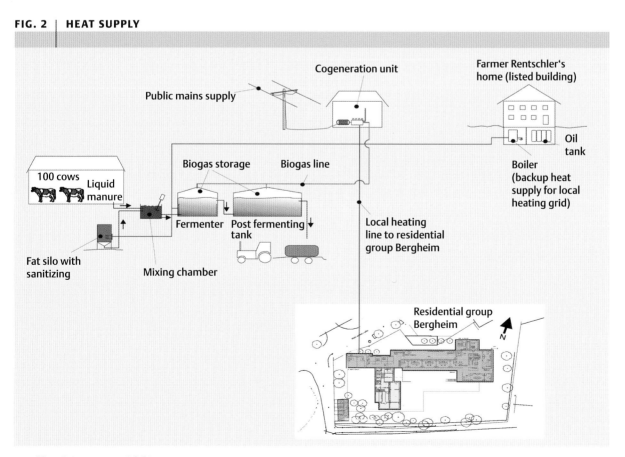

Local heating concept with biogas system.

FIG. 3 | CO₂ BALANCE

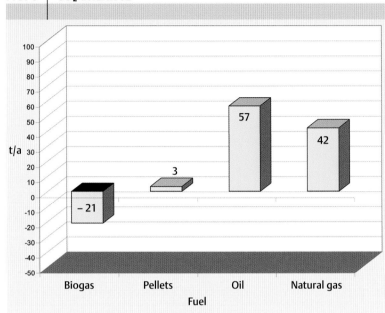

Balances of annual greenhouse gas emissions, converted to tons of saved CO₂, for selected heating concepts in buildings. The –21 tons corresponds to the methane that is no longer released by the produced liquid manure in the farm due to the biogas system.

ed to 161 megawatt hours per year (MWh/a). Here, even after renovating, the consumption of heating energy in the old building is twice as high as in the new building.

For assessing potential greenhouse gas emissions, all supply variants such as biogas, pellets, as well as natural gas and oil were investigated in terms of their CO₂ emissions (Figure 3). The biogas was by far the best alternative: moreover, it even features negative CO₂ potential, i.e., it releases less greenhouse gas compared to the situation if the therapy center had not been built at all. As mentioned, this startling result is due to the fact that without the biogas system, the farmer's livestock would release methane into the atmosphere, which now is not the case. This reduces methane emissions noticeably, and corresponds to savings of 21 tons of CO₂ per year. Compared to the unfavorable case of integrated oil heating, the design yields additional savings of 57 tons of CO₂ per year.

The concept realized now thus saves 78 tons of CO₂ per year, compared to a conventional building. This clearly shows how effectively such a sustainable neighborhood solution can protect the environment and simultaneously provide an autonomous energy supply. It is a good example of how problems are better solved in cooperation than by working alone.

Summary

The example of a therapy building for approximately 40 residents shows the advantages of an intelligent overall concept. The complex comprises an old, renovated villa and an energy-efficient, new building. Large windows combined with good insulation use solar heat in winter and reduce electricity-consuming artificial lighting. Via an earth duct, the ventilation system beneath the new building preheats the fresh air in winter and cools it in summer. A neighboring farmer provides heating and hot water. He built a biogas system with co-generation unit that converts the liquid manure of his cows and other organic waste into renewable electricity. The farmer feeds this electricity into the grid. This concept reduces the total greenhouse gas emissions of the building complex including the farmhouse tremendously. Thanks to the subsidies in Germany, it has already become cost effective.

About the Authors

Matthias Schuler, born 1958, studied mechanical engineering at the University of Stuttgart, Germany, focusing on technologies for efficient energy useage. In 1992, he founded Transsolar, Stuttgart, and since then has been head of the firm. He teaches at Biberach University of Applied Sciences and at Stuttgart University. Since 2001, he has been lecturer in architecture at Harvard University Cambridge, MA.

Christian Matt, born 1962, studied mechanical engineering at the University of Stuttgart, Germany. In his diploma thesis at the Institute of Thermodynamics, he investigated a thermal (solar) powered air ventilation system. Since 1996, he has been working for Transsolar, and he is project manager since 2002. Projects focus on concepts for building renovation and sustainably-regeneratively supplied buildings.

Contact:
Christian Matt,
Transsolar Energietechnik GmbH,
Curiestraße 2, D-70563 Stuttgart, Germany.
matt@transsolar.com

Promotion of Renewable Energy in Germany

How Political Will Changes a Country

Thomas Bührke

Promoting renewable energy, saving energy, and reducing greenhouse gas emissions – for a number of years, these have been the goals of the German Federal Government.

In July 2005, the German Federal Government passed the National Climate Protection Program. In this document, the government pledges itself to cut greenhouse gas emissions by 21 % of their 1990 level, within the period from 2008 to 2012. Until 2020, they are to be reduced by 40 % if the EU decides on an emission goal of 30 %. Moreover, for the year 2050, a reduction of 80 % is aspired to.

For reaching these goals, the German Federal Government has taken a large number of measures. These include the Renewable Energy Sources Act (EEG), development of cogeneration, the Energy Saving Ordinance (EnEV), improvement of rail traffic, etc.

The energy sector plays a vital role in climate protection since it alone releases approximately 40 % of the national CO_2 emissions. This is why the German Federal Government has triggered an initiative for promoting renewable energy. The goal is to double their share of the total energy supply until 2010, compared to the portion in 2000. For electricity, this means an increase to 12.5 %, and 4.2 % for primary energy consumption. By the end of 2006, this goal had practically been achieved. 12 % of the electricity and 5.8 % of consumed primary energy in Germany originated from regenerative sources, a 2007 publication of the Federal Ministry for the Environment, Nature Conservation, and Nuclear Safety says. It is planned to increase the share of renewable energy in electrical power supply to at least 20 % by 2020 and to approximately 50 % by the mid 21st century.

The EEG is a vital component of energy policy. This act obliges electrical energy suppliers to accept delivery of electricity from solar power, hydropower, wind, geothermal power, as well as biomass and to pay certain minimum fees. On August 1, 2004, the amendment of the EEG became effective. This improved in particular the general framework for input, transfer, and distribution of electrical power from renewable energy sources. Support programs for energy from wind, solar power, biomass, and geothermal systems initiated by the German Federal Government have additionally fostered the necessary speed-up of the regenerative energy sources' market launches. These programs are supplemented with a large number of subsidizing programs implemented by the European Union, federal states, local authorities, and energy suppliers.

Consistent energy savings can also promote using less fuel, thereby cutting carbon dioxide emissions. Therefore, the building sector is one of the most important fields of activity in terms of climate protection. For instance, 78 % of the final energy consumption in German private households – not including transportation – is spent on space heating. Another 10 to 15 % are used for water heating. Potential savings are tremendous: existing homes, on average, consume nearly three times the energy tolerable for new buildings according to the requirements of the new Energy Saving Ordinance (EnEV).

Thus, since 2003, the German Federal Government has added 160 million euros (approx. 230 million dollars) annually to the CO_2 building reconstruction program of the federal KfW banking group for older buildings, raised by the ecological tax and financial reform, thus nearly doubling the support to approximately 360 million euros (520 million dollars) today (KfW: Kreditanstalt für Wiederaufbau; the KFW banking group is owned by the German federal government and partly by the states of Germany. It was founded in 1948 in Western Germany to finance the reconstuction of the country).

In 2006 and 2007, the financial volume available in Germany for continuing the program amounted to a total of 720 million euros (more than 1 trillion dollars), allocated to interest rate reduction and partial debt relief.

About the Author

Thomas Bührke, postdoctoral physicist, is an editor of the German physics journal "Physik in unserer Zeit" and a freelance science journalist.

Contact:
Dr. Thomas Bührke,
Physik in unserer Zeit,
Wiesenblättchen 12,
D-68723 Schwetzingen, Germany.

Renewable Energy. Edited by R. Wengenmayr, Th. Bührke. Copyright © 2008 WILEY-VCH Verlag GmbH & Co. KGaA, Weinheim. ISBN 978-3-527-40804-7

Subject Index

Renewable Energy. Edited by R. Wengenmayr, Th. Bührke. Copyright © 2008 WILEY-VCH Verlag GmbH & Co. KGaA, Weinheim. ISBN 978-3-527-40804-7

There are
more efficient ways
to find hidden
information.